Lecture Notes in Economics and Mathematical Systems   663

W0235500

For further volumes:
http://www.springer.com/series/300

Svenja Lagershausen

Performance Analysis of Closed Queueing Networks

 Springer

Svenja Lagershausen
Department of Supply Chain
Management and Production
University of Cologne
Cologne, Germany

ISSN 0075-8442
ISBN 978-3-642-32213-6   ISBN 978-3-642-32214-3 (eBook)
DOI 10.1007/978-3-642-32214-3
Springer Heidelberg New York Dordrecht London

Library of Congress Control Number: 2012950583

Printed on acid-free paper

Springer is part of Springer Science+Business Media (www.springer.com)

Acknowledgment

This Ph.D. thesis would not exist if it wasn't for many people who helped me with their professional advice, emotional support, and also by keeping me from work. First of all, I would like to thank my supervisor, Prof. Dr. Horst Tempelmeier for having me as his research assistant and for his great support and advice throughout my time in his department. I also thank Prof. Dr. Stefan Helber, Prof. Dr. Michael Manitz, and Prof. Dr. Raik Stolletz for having an "open ear" and giving me lots of tips and advices for my studies. I am thankful to Prof. Dr. Ulrich Thonemann for being co-referee of my Ph.D. thesis. I thank Dr. Johannes Antweiler for his great support both in good and bad times - it has made a great impact. I am also thankful to Manuela Pioch for her help in organizational matters and for being very supportive. I would like to thank my colleagues Dipl.-Wi.-Inf. Dr. Sascha Herpers, Dipl.-Wi.-Inf. Karina Copil, and Dipl.-Kfm. Oliver Bantel for proofreading my manuscript, for discussing all kinds of problems and, most important, for being such good friends. To all people of the department goes a big thank you for a great atmosphere which made the department a joyful place to work in. I also thank Julie Teft, Sigrid Newman, Sebastian Doblhofer, Katrin Schmitter, and Jonas Strack for their proofreading. Further, I thank Desdemona Mller, Christiane Degen, Verena Pick, Birgit Schorn, Anna-Lena Leifert, Jan Hendrik, Miriam Schmidt, Andr Fuetterer, Raik zsen, Dirk Briskorn, Raik Stolletz, Michael Manitz, Timo Hilger, Sascha Herpers, Karina Copil, Oliver Bantel, Duc Hung Tran, and Aurelia Froitzheim for their friendship and the wonderful time in Cologne and in different places throughout Europe! Finally, I thank my parents for financially enabling me to study in the first place and by this laying the foundation for my professional life. Also, I thank my Mom, Dad, sister Wiwi, and grandfather for their great emotional support and interest throughout the years and for just being there!

THANK YOU ALL!
Cologne, July 2012.

About the author

Svenja Lagershausen is employed as a research assistant at the Leibniz University of Hanover, Germany, at the Department of Production Management. Prior to her current position, she worked at the Department of Supply Chain Management and Production, University of Cologne. During this time, she wrote on her Ph.D. thesis "Performance Analysis of Closed Queueing Networks." Her research interests are in the field of queueing theory, design and operation of automated manufacturing systems, and production planning. Further work from her has been published in journals as Annals of Operations Research and International Journal of Production Research.

Contents

List of Figures

List of Tables

List of Algorithms

List of Symbols

$\beta_i(n_i)$	helping term for the calculation of $G(N)$
γ_i	workload at station i, $\gamma_i = \frac{e_i}{\mu_i}$
λ	virtual arrival rate to an open queueing network that is equivalent to the CQN being considered
$\lambda_{\max}$	upper bound for λ
$\lambda_{\min}$	lower bound for λ
$\lambda_c^{(i)}(n-n_i)$	load-dependent arrival rate of $n-n_i$ customers at the composite node c
$\lambda_i(n_i)$	load-dependent arrival rate to station i when there are n_i customers at station i
$\lambda_{\mathcal{N}_i}(u)$	exponential rate of non-active state u regarding station i
$\mathcal{A}_i^F(ph)$	set of active states from which the workpiece at station i is potentially finished in the ph-th phase, $\mathcal{A}_i^F(ph) \subseteq \mathcal{A}_i$
$\mathcal{A}_i^{FN}(ph)$	set of potential transition-into-non-activity states in which station i potentially finishes after the ph-th phase, $\mathcal{A}_i^{FN}(ph) \subseteq \mathcal{A}_i^F(ph)$
$\mathcal{A}_i$	set of processing states regarding station i
$\mathcal{N}_i^E$	set of non-active states that can be entered from an active state of station i, $\mathcal{N}_i^E \subseteq \mathcal{N}_i$
$\mathcal{N}_i^N$	set of non-active states which can only be entered by another non-active state regarding station i, $\mathcal{N}_i^N \subseteq \mathcal{N}_i$
$\mathcal{N}_i$	set of non-active states regarding station i with $\mathcal{N}_i = \{\mathcal{N}_i^E, \mathcal{N}_i^N\}$
$\mathcal{P}_i^E$	set of exits of the phase-type processing time distribution at station i
$\mathcal{P}_i$	set of processing phases at station i
$\mathcal{P}_i^E$	set of processing phases at station i with exit possibility
$\mathcal{P}_i^N$	set of processing phases at station i without exit possibility
$\mathcal{P}_j$	set of the predecessor stations of work station j

$\mathcal{S}$	state space of the Markov chain
$\mathcal{S}_i$	set of the successor stations of work station i; $\mathcal{S}_M$ is the set of input stations that receive production-authorization information from the final work station M. With our assumptions, it holds that $\mathcal{S}_M = \{1\}$
$\mu_d(i, j)$	virtual service rate in the subsystem (i, j)
μ_i	mean processing rate at station i
μ_p	exponential processing rate of the p-th phase
$\mu_u(i, j)$	virtual arrival rate to the subsystem (i, j)
μ_{i, ph_i}	exponential rate at station i in phase ph_i
μ_{sti}^τ	transition rate of the transition from state s to state t induced by station i
$\nu_i(n_i)$	load-dependent departure rate of station i when there are n_i customers at station i
$\overrightarrow{bp}$	vector indicating if a station can potentially be blocked, $bp_i = 0$ if station i cannot be blocked and $bp_i = 1$ if station i can be blocked
$\overrightarrow{bs}$	binary vector indicating which stations are blocked with $bs_i = 0$ if station i is not blocked and $bs_i = 1$ otherwise
$\overrightarrow{bs}^I$	initial blocking statuses, $bs_i^I = 0 \ \forall i$
$\overrightarrow{b}$	buffer capacity over all stations
$\overrightarrow{ph}$	vector of active-phase indices over all stations
$\overrightarrow{ph}^I$	initial active phases, $ph_i^I = 1$ if $w_i \geq 1$ and $ph_i^I = 0$ if $w_i = 0$
$\overrightarrow{ph}^{LB}$	vector of lower bounds of the active-phase indices at all stations
$\overrightarrow{ph}^{NR}$	number of phases according to the phase-type distribution
$\overrightarrow{ph}^{UB}$	vector of upper bounds of the active-phase indices at all stations
$\overrightarrow{q}$	queue lengths at all stations, $q_i = \max\{w_i - 1, 0\}$
$\overrightarrow{w}$	workpiece allocation over all stations
$\overrightarrow{w}^I$	initial workpiece allocation
$\pi(n_1, \ldots, n_M)$	marginal probability for the distribution of customers over the stations $(n_1, \ldots, n_M)$
$\pi(s)$	steady-state probability of residing in state s
$\pi_{\mathcal{N}T_i}(u)$	probability of transiting into a particular non-active state u after the finish of a workpiece at station i
$\pi_{\mathcal{N}_i}$	probability of transition from activity to non-activity regarding station i
$\pi_{T_i}(ph, u)$	probability of transiting into a particular non-active state u from an active state regarding station i
σ	vector of the initial distribution
$\zeta_d(i, j)$	coefficient of variation of the holding time at the downstream station in subsystem (i, j)

$\zeta_u(i,j)$	coefficient of variation of the virtual inter-arrival time at the upstream station in subsystem (i,j)
a_i	probability of continuing the phase-type distribution at station i after the first phase
$afn(ph,u)$	active state that leads into non-active state u when finishing the workpiece after phase ph, $afn(ph,u) \in \mathcal{A}_i^{FN}$
b_i	buffer capacity at station i
bs_{si}	value of blocking statuses at station i in state s
c_i	coefficient of variation of the processing time distribution of station i
D	time between departures
d^{min}	capacity of the station with least buffer and server capacity, $d^{min} = \min_i\{b_i + 1\}$
d_i	upper bound of the number of workpieces at station i, $d_i = \min\{n, b_i + 1\}$
$E[T_i]$	expected value of the processing time at station i
e_i	visiting ratio of station i
f	highest station index for which the workpiece allocation is fixed (starting from the first station, i.e. $i = 1,\ldots,f$)
$F_i(n_i)$	weighting term for the calculation of $G(N)$
$G(N)$	normalizing constant
g_i	station capacity, $g_i = b_i + 1$
i^{UB}	lowest station index containing a workpiece
k	number of phases in the phase-type distribution
$L(i,j)$	expected work-in-process in the subsystem (i,j) of the equivalent open queueing network
L_i^Q	mean queue length at station i
L_i^S	mean work-in-process at station i
L_i^Q	mean number of jobs in the queue of station i
M	number of stations in the system
N	maximum number of workpieces in the CQN
n	number of workpieces/customers/jobs/pallets circling in the system
n^R	number of remaining workpieces, $n^R = n - \sum_{i=1}^{f} w_i$
n^{NS}	minimum number of workpieces for which no starving occurs in the complete networks
n_i	number of customers at station i
NRT_i	number of exits after the first phase of station i
nrt_i	index of exit after the first phase of station i

$o(h)$	any function converging faster to zero than its argument, $\lim\limits_{h \to 0} \frac{o(h)}{h} = 0$
$p'_{ij}(t)$	marginal probability of transiting from state i to state j during time period t
$P(n_1, \ldots, n_M)$	probability of the customer allocation $n_1, \ldots, n_M$
$p^*_B(i, j)$	arrival instant blocking probability in subsystem (i, j)
$p^*_S(i, j)$	arrival instant starving probability in subsystem (i, j)
$P_i^B(n)$	probability of blocking at station i if n workpieces circulate in the system
$P_i^S(n)$	probability of starving at station i if n workpieces circulate in the system
$p_i^{TBPS}(d, e)$	probability of transiting from state d into state e, where d and e are states of the $TBPS_i$-distribution
$p_B(i, j)$	time-average blocking probability in subsystem (i, j)
$P_i(n_i)$	steady-state probability of n_i workpieces at station i
P_i^B	blocking probability at station i
$p_N(u, v)$	transition probability from non-active state u to non-active state v
$p_N(u, v)$	probability of transiting from state u to state v, where u and v are elements of the set of non-active states $\mathcal{N}_i$
$p_S(i, j)$	time-average starving probability in subsystem (i, j)
p_{ij}	probability of transiting from state i to state j
$p_{ij}(0, h)$	probability of transiting from state i to state j in the time interval $(0, h]$
$p_{ij}(h)$	probability of transiting from state i to state j during time period h
ph_i^{LB}	lower bound of the active-phase index at station i
ph_i^{UB}	upper bound of the active-phase index at station i
ph_i	index of active phase at station i, $ph_i = \mathbb{1}_{\{n_i \geq 1\}}$
ph_i^C	value of phases at station i in the current realization with $ph_i^C = 1, \ldots, ph_i^{NR}$ if $w_i \geq 1$ and $ph_i^C = 0$ if $w_i = 0$
ph_i^{NR}	number of phases of the phase-type distribution at station i
ph_{si}	index of the active phases at station i in state s
$PR(n)$	production rate with n workpieces in the system
Q	transition rate matrix
$Q(s, t)$	transition rate from state s to state t
Q_i^{TBPS}	transition rate matrix of the $TBPS_i$-distribution regarding station i
q_i	queue length at station i, $q_i = \max\{n_i - 1, 0\}$
q_{ij}	rate of transiting from state i to state j
q_i	rate of leaving state i
q_{si}	queue length at station i in state s

S	matrix of transition rates between states; part of matrix of Q
s	index of a state in the Markov chain
S^0	vector of the transition rates into the absorbing state; part of matrix of Q
s_i	server capacity of station i
T_i^B	mean blocking time at station i
T_i^Q	mean waiting time in the buffer in front of station i
T^S	cycle time in the system
T_i	random variable representing the processing time at station i
T_i^S	cycle time at station i
T_i^{Stv}	mean starving time at station i
$tr_i^{TBPS}(d,e)$	transition rate from state d into state e, where d and e are states of the $TBPS_i$-distribution
$TBPS_i$	time between processing starts at station i
U_i	utilization at station i
$Var[T_i]$	variance of the processing time
w_i	number of workpieces at station i
$WIP(\lambda)$	Expected work-in-process in the open queueing network for a given virtual arrival rate λ
$X(\lambda)$	Production rate of the equivalent open queueing network for a given virtual arrival rate λ
$X(t)$	random variable of a stochastic process at time t

List of Abbreviations

AMVA	approximate mean value analysis
ARW	allocation of remaining workpieces
BAS	blocking after service
BBS	blocking before service
CCNC	coalesce computation of normalizing constants
CONWIP	constant work in process
CQN	closed queueing network
CV	coefficient of variation
EBOTT	extended Bottapprox-method
ESCAT	Extended Self-Correcting Approximation Technique
ESUM	Extended Summation Method
FCFS	first-come first-served
FMS	flexible manufacturing systems
GMRES	generalized minimal residual method
HAM	heuristic aggregation method
LBANC	local balance algorithm for normalizing constants
MVA	Mean value analysis
MVAQ	Mean value analysis of queues
PFS	product-form solution
SCAT	Self-Correcting Approximation Technique
SUM	summation method

Chapter 1
Introduction

1.1 Motivation

In many automated production systems, carriers are used to transport the work-pieces. Since these carriers circulate through the stations, the production systems can be modeled as closed queueing networks (CQN).

The processing times at the work stations are often stochastic. Reasons for the stochastics are, for example, manual operations, rework, or a variety of products sharing a production line. Furthermore, stochastic failures of machines may be incorporated into the processing times, thus leading to effective stochastic processing times. In order to be able to model the processing time close to the empirically estimated distribution, the first and the second moment should be provided. Such distributions are here referred to as general distributions.

Because of the stochastic processing times, buffers are needed between the stations in order to reach a predefined production rate. However, the installation of buffers is a significant financial investment. Furthermore, additional buffer capacity leads to a higher level of work-in-process, which ties up capital. As a result, buffer optimization is an important issue. In order to find the optimal configuration of a network, it is necessary to evaluate a high number of configurations. Therefore, the requirement on performance-analysis procedures is that they are fast, provided that the results are correct. The most important measure constitutes the production rate, but also other measures such as the mean waiting time, the mean cycle time, and the mean queue length are of interest.

Methods used for the performance analysis can be classified into simulative and analytical methods. The latter may be further divided into exact and approximate methods. The advantage of simulation is that the network can be precisely and flexibly modeled. However, a long computation time is required until statistically significant results are achieved. As a consequence, one is forced to restrict the number of configurations that can be analyzed during the optimization process,

S. Lagershausen, *Performance Analysis of Closed Queueing Networks*, Lecture Notes in Economics and Mathematical Systems 663, DOI 10.1007/978-3-642-32214-3_1,
© Springer-Verlag Berlin Heidelberg 2013

which often leads to suboptimal decisions.[1] In contrast, analytical methods provide performance-measure estimates quickly, and thus, allow the evaluation of a high number of configurations.

Exact methods for the analysis of CQN are limited to systems with exponential distributions and limited buffer capacities, as well as to systems with general processing times and unlimited buffer capacities where the system consists of more than two stations. For two-station systems, however, an exact analysis is possible under the assumption of general processing times and limited buffers. Furthermore, there are approximate analytical approaches under the assumption of general processing times and finite buffer capacities.

In this thesis, both an exact and an approximate analytical method are proposed for the calculation of the production rate and other performance measures. Moreover, an exact approach for the distribution of the time between processing starts is presented. Throughout the thesis, closed queueing networks with a linear flow of material, general processing times, and finite buffer spaces are considered. Further assumptions are a blocking-after-service discipline, a first-come first-served queueing discipline, single-server stations, and one customer class.

1.2 Structure

The content is structured as follows. In Chap. 2, the assumptions are introduced, and the characteristics of the CQN under consideration are presented. The literature is reviewed in Chap. 3; it contains exact and approximate methods for the performance measures of closed queueing networks with exponential or general processing times, as well as with unlimited or limited buffer capacities.

In Chap. 4, an approximate decomposition approach is proposed based on an existing procedure. The proposed method is fast and accurate compared to existing ones. In this procedure, a virtual arrival rate to an open queueing network is to be found such that the specified work-in-process corresponds to the given number of carriers. For this purpose, an estimate of the work-in-process in G/G/1/K systems is introduced.

Subsequently, in Chap. 5, the exact performance analysis by Markov chains is presented. The processing times are assumed to follow phase-type distributions. The entire closed queueing network is modeled as one Markov chain, considering the blocking states and transitions according to the specified phase-type distribution. The derivation of the global balance equations is presented in order to explain the theoretical background of the Markov-chain analysis. Further, the implementation of the approach, which allows an automated evaluation of an arbitrary number of configurations, is described. In a computational study, the runtime performance is investigated and numerical results are provided.

---

[1] See Manitz (2005, p. 2).

Lastly, in Chap. 6, a method for the exact distribution of the time between processing starts is presented. This distribution is of interest for decisions concerning actions that take place at the instant of a processing-start. It is based on the Markov-chain approach proposed in the preceding chapter. The distribution results in a general phase-type distribution which is composed of the processing time distribution and the blocking and starving states of the considered station. The influences of the configuration parameters on the coefficient of variation of the time between processing starts are analyzed by means of a numerical study.

Chapter 2
Closed Queueing Networks

Queueing networks in general are networks of processing stations with intermediate storage areas called buffers. At each of the stations, a service is provided that takes up time. Workpieces approach the stations and are processed immediately if the server is idle. Otherwise, the workpieces line up in the buffer in front of a station and wait to receive service. Apart from that, queueing networks differ.

In Sect. 2.1, we present the assumptions of the closed queueing networks considered in this thesis and contrast these to other common assumptions. Subsequently, in Sect. 2.2, we introduce the characteristics of the closed queueing networks in focus.

2.1 Assumptions

Material flow. With regard to the material flow, queueing networks are mainly distinguished between open queueing networks (OQN) and closed queueing networks (CQN). Open systems are characterized by an arrival process which is independent of the departure process. In these systems, the first station is never starved and the last station is never blocked.[1]

In closed queueing networks, workpieces circulate through the system. The arrival stream at the first station conforms with the departure process at the last station. In contrast to open queueing systems, the last station of a closed queueing network may become blocked if the buffer in front of the input station is full of workpieces. This holds true under the assumption that the buffer capacities within the system are finite. Furthermore, the first station may become starved if no carriers with workpieces are available in the buffer in front of the input station. Closed systems, therefore, are characterized by high dependencies not only between

---

[1] See Dallery and Gershwin (1992).

S. Lagershausen, *Performance Analysis of Closed Queueing Networks*, Lecture Notes in Economics and Mathematical Systems 663, DOI 10.1007/978-3-642-32214-3_2, © Springer-Verlag Berlin Heidelberg 2013

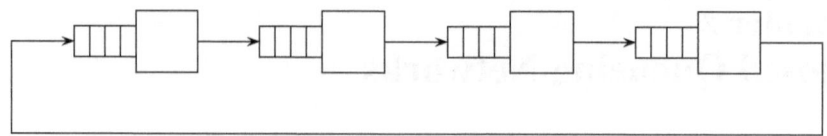

Fig. 2.1 Closed queueing network with four stations

Table 2.1 Notation

b_i	Buffer capacity at station i
c_i	Coefficient of variation of the processing time distribution of station i
d_i	Capacity of the station i, $d_i = b_i + s_i$
d^{min}	Capacity of the station with least buffer and server capacity, $d^{min} = \min_i\{b_i + 1\}$
M	Number of stations in the system
μ_i	Mean processing rate at station i
n	Number of workpieces/customers/jobs/pallets circling in the system
n^{NS}	Minimum number of workpieces for which no starving occurs in the complete networks
N	Maximum number of workpieces in the CQN
$P_i^B(n)$	Probability of blocking at station i if n workpieces circulate in the system
$P_i^S(n)$	Probability of starving at station i if n workpieces circulate in the system
$PR(n)$	Production rate with n workpieces in the system
s_i	Server capacity of station i
T_i	Random variable representing the processing time at station i

mid-stations, but also between the first and the last station. The closed-loop flow reflects the main assumption regarding the queueing networks examined in this thesis.

An example of a closed queueing network with four stations and linear flow of material is depicted in Fig. 2.1. Each station consists of a server (taller rectangle) and buffer space in front of the server (smaller rectangles).

Within the class of closed queueing networks, there are systems with a linear flow of material, flexible manufacturing systems,[2] systems with arbitrary routing,[3] and closed assembly and disassembly systems.[4] Here, the flow of material is assumed to be linear, i.e. the processing stations are connected in series. In this case, each workpiece receives service in the same order. We denote the stations by the index i in topological order, ranging from 1 to M, with M denoting the number of stations. In closed systems, the successor of station M constitutes station 1, and vice versa, the predecessor of station 1 represents station M. The notation used in this chapter is given in Table 2.1.

Work-in-process. The material consists of discrete parts. The number of items circulating within the closed queueing network is constant. This constitutes the central assumption in closed queueing networks. The circular flow of material is

---

[2]See Tempelmeier and Kuhn (1993).

[3]See Koenigsberg (1982).

[4]See Duenyas (1994).

often due to the requirement that workpieces must be led through the production system by carriers. In this case, a raw material is attached to an empty carrier when it passes in front of the first station. The carrier leads the workpiece through the system. Behind the last station, the finished product is released, and the empty carrier picks up raw material anew in the buffer between the last and the first station. No carriers are added or removed causing the number of carriers to stay constant.

Alternatively, the circulating items can be production-authorization cards, also called CONWIP (constant work-in-process) cards. In a CONWIP system, cards are attached to each processing unit. The purpose of these cards is to control the work-in-process of the system: After the completion of a unit at the last station, the CONWIP card is released from the finished workpiece. This free card authorizes a raw unit to enter the system in front of the first station.

We assume that, at the instant at which a finished product is removed from the carrier (or card), a new workpiece is loaded onto that carrier (or attached to the card) infinitely fast. In other words, the time to change a finished product into a raw product amounts to zero. As a result, not only are the carriers or cards constant, but so is the work-in-process.

The number of carriers or workpieces is denoted by n. n ranges from 1 to N, where N denotes the system capacity minus one.[5] The number of carriers greatly influences the production rate. This issue is investigated in Sect. 2.2.

Processing time distribution. We assume that the processing time at station i, denoted by T_i, represents a random variable describing the time a server needs for the processing of a workpiece. T is assumed to be independent and identically distributed for all workpieces. The processing time per station may vary because of products taking different amounts of time or due to manual labor. Moreover, machine failures may contribute to the variability of the processing time if the machine failures are implicitly considered as part of the processing time per workpiece.[6]

The most frequently used processing time distribution is the exponential distribution. Systems under this assumption are mathematically easier to handle. However, the exponential distribution implies a very high variability, which is usually not present in real-life systems.[7]

In order to model the processing time more accurately, the variance should be regarded as well. We assume the processing time to be specified by the mean processing rate, denoted by μ, and the coefficient of variation, denoted by c (both of arbitrary values),[8] or that it follows a phase-type distribution.[9] Both settings also include the exponential distribution.

---

[5]For the range of the number of workpieces, see also page 9.

[6]See Gaver (1962) and "Machine failures and repairs" in this section.

[7]See Sect. 5.2.2 for further details.

[8]This is assumed in Chap. 4.

[9]In Chap. 5, both cases are considered.

If the processing times are stochastic, the so-called starving effect occurs. Starvation takes place if a station is operative but not supplied with material. It is measured by the starving probability that corresponds to the percentage of time in which starving occurs. It holds that the higher the starving probability, the lower the production rate.

Buffer capacity. Under the assumption of stochastic processing times, buffer space between the stations is very important because it mainly contributes to the productivity of the network. Buffer capacity ranges from infinitely large (unlimited buffer capacity), over a defined number (limited buffer capacity), to no buffer capacity. Unlimited buffer capacity is a theoretical construct that is easier to handle in performance-analysis procedures. Procedures for limited buffer-capacity systems are able to consider the complete range of buffer capacity from infinite to nonexistent. Within this thesis, limited buffer capacity is assumed. The capacity of the buffer in front of station i is denoted by b_i.

If the buffer space is limited, blocking may occur. A station is blocked if it is unable to work because the succeeding buffer is full and is, therefore, prevented from working on the next job. The higher the buffer capacity, the lower the frequency of blockages, which results in a higher production rate.

Although higher buffer capacity increases the production rate, it is not beneficial to install as many buffers as possible. In buffer optimization, the buffer capacity and the buffer distribution constitute important decision variables in the optimization of queueing networks and production systems in particular.[10] In closed queueing systems, the number of workpieces represents a further decision variable in the optimization.

Blocking mechanism. Blocking may occur at different points during operation. The two most common blocking mechanisms are blocking-after-service and blocking-before-service.[11] This thesis is based on the blocking-after-service discipline.

Under the blocking-before-service (BBS) mechanism—also called type-2 blocking or service blocking—a machine can only start processing if a space is available in the downstream buffer. This means that a station is blocked if the succeeding buffer is full at the instant that the processing is supposed to start.

Blocking-after-service (BAS)—also called type-1 blocking or production blocking —occurs if, at the instant of completion of a part, the downstream buffer is full. The finished workpiece stays on the machine and prevents the machine from further production—thus the server is blocked. As soon as the downstream machine releases its current workpiece, it starts processing the next workpiece and makes buffer space available. This is the instant in time, in which blocking is resolved.

---

[10] See for instance Gershwin and Schor (2000) on a buffer optimization procedure.

[11] See Dallery and Gershwin (1992, p. 12), for these and other blocking mechanisms.

Server. A network is distinguished by single and multiple servers. We assume that the processing unit consists of a single server. That means only one workpiece may be on the server. The three states of the processing unit are starved, blocked, and busy.

Machine failures and repairs. In some systems, machines are prone to failure. There are two major types of failures described in the relevant literature: operation-dependent failures and time-dependent failures.[12] In this thesis, machine failures are not considered explicitly, i. e. the servers are assumed to be completely reliable. However, operation-dependent failures may be implicitly included in the processing times by the Completion-Time Concept of Gaver (1962). This concept is explained in detail in Manitz (2005).[13]

In summary, our assumptions are as follows: The flow of material is assumed to be linear. At each of the $i = 1, \ldots, M$ stations, a single server operates with stochastic processing times without failures. The service time distribution is described by the processing rate μ_i and the coefficient of variation c_i at each station i or by a phase-type distribution. In front of each station, a buffer with finite capacity b_i is located. Processing takes place according to the first-come first-served service discipline. After the process completion, the workpieces are led into the subsequent buffer if space is available. Otherwise, the server is blocked. The blocking mechanism is assumed to be blocking-after-service (BAS). Behind the last station, finished products are released and empty carriers pick up raw material in the buffer in front of the first station.

2.2   Characteristics

The defining characteristic of closed queueing networks is the constant number of workpieces. This mainly influences the performance measures of the system. The production-rate function subject to the number of workpieces is characteristic for CQN. We will, therefore, investigate the effect of the number of workpieces on the blocking and starving probabilities leading to the typical production-rate function according to the assumptions made above.

Range of the number of workpieces. The number of workpieces ranges from 1 to N. It is restricted to N due to the finite capacity of the system. If the number of workpieces amounted to the number of places in the system, each server would be blocked by the parts in the succeeding buffer, and the production rate would amount to zero. This effect is called deadlock.[14] Therefore, n must not exceed one workpiece less than the sum of the buffer and server capacities, b_i and s_i, of all stations i.

---

[12]For further details, see Dallery and Gershwin (1992, pp. 14ff).

[13]See Manitz (2005, pp. 35ff).

[14]See Yüzükirmizi (2005, pp. 17f).

The maximum number of workpieces, N, is given by

$$N = \sum_{i=1}^{M} (b_i + s_i) - 1. \tag{2.1}$$

Blocking. Blocking is measured by the blocking probability which corresponds to the percentage of time in which station i is blocked. It is denoted by P_i^B. As long as the number of workpieces in the system, n, is less or equal to the minimum station capacity, d^{min}, with $d^{min} = \min_i\{b_i + s_i\}$, no blocking can occur at any station:

$$P_i^B(n) = 0 \qquad \text{for } n \leq d^{min}, \forall i. \tag{2.2}$$

With one more workpiece, $n = d^{min} + 1$, blocking can occur at the station located upstream of the minimum-capacity station. We denote the index of the station with the minimum station capacity by j. If, at any instant in time, the number of workpieces at station j equals $n_j = d^{min}$ and the remaining workpiece is located at station $j - 1$, $n_{j-1} = 1$, then station $j - 1$ becomes blocked if it finishes its current workpiece faster than station j.[15] Generally, station i may be blocked as soon as the number of workpieces in the system is greater than the station capacity of the succeeding station, $d_{i+1} = b_{i+1} + s_{i+1}$:

$$P_i^B(n) > 0 \qquad \text{for } n > d_{i+1}, \forall i. \tag{2.3}$$

In the performance analysis, blocking must be taken into account as soon as the first station might become blocked, i. e. if $n > d^{min}$.

Starving. Starving is expressed by the starving probability, which represents the percentage of time in which a station is not supplied with material. This probability is denoted by P_i^S with regard to station i. With only one workpiece in the system, the starving probability is as high as possible. By adding more workpieces to the system, the probability of starvation decreases more and more.

Upon reaching a high enough quantity of workpieces, station i cannot be starving anymore, and the server of station i is occupied at any time. This occurs at a workpiece level such that, even if all buffer places and servers of all other stations are occupied, at least one workpiece still remains to be at station i:

$$P_i^S(n) = 0 \qquad \text{for } n > \sum_{\substack{k=1, \\ k \neq i}}^{M} (b_k + s_k) \ \forall i. \tag{2.4}$$

---

[15]See Onvural and Perros (1989b, p. 112).

Table 2.2 Exemplary configuration of a CQN

i	1	2	3	4	5
μ_i	1	1	1	1	1
c_i^2	0.5	0.6	0.5	0.6	0.7
b_i	4	4	4	4	4
s_i	1	1	1	1	1

Starving occurs if less than the afore-described number of workpieces resides at the station:

$$P_i^S(n) > 0 \qquad \text{for } n \leq \sum_{\substack{k=1, \\ k \neq i}}^{M} (b_k + s_k) \ \forall i. \qquad (2.5)$$

There is no starving in the complete network if the number of workpieces is higher than n^{NS}, where n^{NS} denotes the system capacity of all stations except for the station with the minimum capacity:

$$n > n^{NS} = \sum_{\substack{i=1, \\ i \neq j}}^{M} (b_i + s_i), \qquad \text{with } j = arg \min_i \{b_i + s_i\}. \qquad (2.6)$$

Production rate function. The production rate is denoted by PR. It constitutes the average number of finished parts per time unit. The production rate is subject to the number of workpieces and corresponds to the processing rate of station i, μ_i, multiplied by the percentage of time the station works at that processing rate, see Eq. (2.7).

$$PR(n) = \mu_i \cdot [1 - P_i^B(n) - P_i^S(n)] \qquad \forall n, i. \qquad (2.7)$$

The percentage of time the station works is called utilization, U_i, and represents the counter-probability of the event that a station cannot work because it is starving or blocked: $U_i = 1 - P_i^B - P_i^S$. The production rate is equal for all stations i. This corresponds to a law called conservation of flow.[16]

In the following, an exemplary configuration is investigated. The data of this example are given in Table 2.2. Figure 2.2 shows the course of blocking and starving probabilities for the given configuration.

The probability of starvation is very high for small n and decreases down to $P^S(n) = 0$ for $n > n^{NS}$. The probability of blocking equals zero for $n \leq d_i \ \forall i$ and increases until N workpieces are reached. Note that in this example, the capacities are equal over all stations. Hence, for each station, the same limits of n hold regarding whether or not blocking or starving occurs.

---

[16] See Dallery and Gershwin (1992, p. 20).

Fig. 2.2 Blocking and
starving probabilities

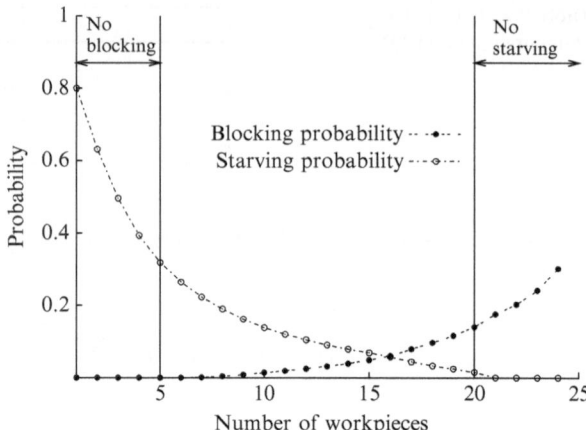

Fig. 2.3 Production rate

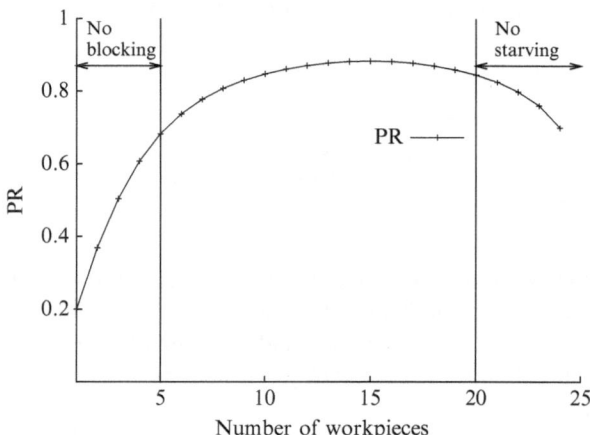

In summary, three ranges can be distinguished. For increasing it n holds that

- For $n \leq d^{min}$, the blocking probability equals zero and the starving probability decreases
- For $n > d^{min}$ and $n < n^{NS}$, both blocking and starving effects occur. The starving probability decreases and the blocking probability increases
- For $n \geq n^{NS}$, the starving probability amounts to zero, and blocking further increases.

The corresponding production-rate function is depicted in Fig. 2.3. The sum of the blocking and starving probability is very high for low n due to the starving probability. It decreases with increasing n because the starving probability decreases. For a high number of n, it increases due to the blocking probability. In accordance with Eq. (2.7), the effect of the production rate is in reverse to the sum of the blocking and starving probabilities: The production-rate function first increases for increasing n and then decreases. This concave function has the typical shape of the production-rate function of CQN.[17]

There is a maximal production rate for a medium number of workpieces. The number of workpieces for which the production rate is maximal, n^*, is found in the range of $d^{min} \leq n^* \leq n^{NS}$. This is true for the following reason: As long as the number of workpieces n is less than d^{min} ($n < d^{min}$), an additional workpiece must increase the production rate at least marginally because no blocking occurs and starvation decreases. As soon as the number of workpieces n exceeds n^{NS} ($n > n^{NS}$), no starvation exists and an additional workpiece will increase blocking and will, therefore, decrease the production rate.

Under optimization aspects, the establishment of a workpiece level n' for which holds $n^* < n'$ is not useful: The same or higher production rate can be achieved with less work-in-process, $PR(n^*) > PR(n')$. A value of n in the range of $1 \leq n \leq d^{min}$ is not of interest from the analytical standpoint because blocking is not taken into account. Hence, the most interesting range of the workpiece level equals $[d^{min} + 1, \ldots, n^*]$.

---

[17]Compare Yao (1985) for statements on closed queueing networks with infinite capacities.

Chapter 3
Literature Review

In recent years, many procedures have been developed to analyze stochastic flow lines. Exact methods and closed-form expressions have been proposed for the analysis of closed queueing networks with exponential processing times and infinite buffer spaces. Under the assumption of exponential processing times and finite buffer spaces, many approximate methods have been introduced. Only a few approximate procedures exist which consider general processing times, of which only a fraction additionally assume finite buffer spaces.

This chapter provides a literature review of performance-analysis methods for closed queueing systems. The focus of the review is on those procedures that treat the problem class defined in Chap. 2. This is the performance analysis of closed queueing networks with a linear flow of material, stochastic processing times without failures, one customer class, and first-come first-served queueing discipline.

Regarding stochastic processing times, analytical methods are distinguished between exponential processing times and general processing times. Concerning buffers, methods are classified into unlimited and limited buffer space. These divisions lead to four different settings. For each setting, the literature is investigated. The settings are:

Setting 1.   Exponential processing times and unlimited buffer capacity
Setting 2.   Exponential processing times and limited buffer capacity
Setting 3.   General processing times and unlimited buffer capacity
Setting 4.   General processing times and limited buffer capacity

The setting of general processing times and limited buffer capacity represents the most general of the introduced settings: The exponential distribution is a special case of a general distribution. Unlimited buffer capacity can be considered by any method for limited buffer capacities, but not vice versa.

Another way to structure evaluation methods is by quality. We differentiate between

Quality 1.   Exact methods and
Quality 2.   Approximate methods

S. Lagershausen, *Performance Analysis of Closed Queueing Networks*, Lecture Notes in Economics and Mathematical Systems 663, DOI 10.1007/978-3-642-32214-3_3, © Springer-Verlag Berlin Heidelberg 2013

Table 3.1 Exact methods (Sect. 3.2)

	Exponential distribution	General distribution
Unlimited capacity	X (Sect. 3.2.1)	X (Sect. 3.2.2)
Limited capacity	X (Sect. 3.2.2)	

Table 3.2 Approximate methods (Sect. 3.3)

	Exponential distribution	General distribution
Unlimited capacity	X (Sect. 3.3.1)	X (Sect. 3.3.3)
Limited capacity	X (Sect. 3.3.2)	X (Sect. 3.3.4)

These classifications lead to eight possible combinations. Tables 3.1 and 3.2 indicate the sections in which the corresponding methods are reviewed. An X shows for which settings and quality methods exist in the literature. Approximate methods are available for each of the settings. However, exact methods do not exist under the assumption of general processing times and limited buffer capacity with more than two stations. In Chap. 5, this gap is closed: An exact method is proposed under the assumption of limited buffer capacity and general processing times.

The defined problem class can be found in the literature among different configurations. These are closed queueing systems, CONWIP-systems,[1] and flexible manufacturing systems. A description of closed queueing networks was provided in Chap. 2.

A "CONWIP-system" refers to a production control mechanism and was proposed by Hopp and Spearman (1990). In this system, CONWIP cards circulate. Behind the last station, the workpiece is released from the system and the card is transferred into the buffer between the first and the last station. In front of the first station, the CONWIP card authorizes the next workpiece to enter the system. For analytical considerations, it does not make a difference whether the work-in-process is thought of as workpiece carriers, authorization cards or customers, it relates to the same problem and results.

Flexible manufacturing systems (FMS) consist of a set of machines, which are connected by an automated transportation system. The system is controlled by a central computer unit. The flow of material is also closed, and stations may be visited in arbitrary order and frequency according to the production plan.[2] Linear CQN represent a special case of FMS. Therefore, procedures for FMS can be used for the analysis of linear CQN as well.

Several literature reviews on closed queueing networks have been proposed. Koenigsberg (1982) presented the development of analytical methods for closed

---

[1] CONWIP = constant work-in-process.

[2] See Tempelmeier and Kuhn (1993, p. 1).

queueing networks from the beginning. Disney and König (1985) reviewed basic concepts regarding queue lengths and cycle times for both open and closed queueing networks. Buzacott and Yao (1986) focused on evaluation methods for flexible manufacturing systems. Onvural (1990) gave a review of closed queueing networks with blocking. Dallery and Gershwin (1992) provided a comprehensive survey on models, properties and features of manufacturing flow-line systems, mainly for open systems. Tempelmeier and Kuhn (1993) gave a review of procedures for performance evaluation and configuration planning of flexible manufacturing systems.

Balsamo, Personè, and Inverardi (2003) carried out a review of open and closed queueing-network models with blocking applied to software architectures. Framinan, González, and Ruiz-Usano (2003) composed a literature review regarding CONWIP-systems. Bolch, Greiner, de Meer, and Trivedi (2006) described many procedures for the performance analysis of queueing networks, especially closed queueing networks. A recent overview of analytical performance-evaluation methods for production systems in general, and for the automotive-industry in particular, was given by Li, Blumenfeld, Huang, and Alden (2009).

3.1 Relevant Theorems

Little's Law

Little's Law, proposed by Little (1961), is a widely applicable law, which is used in many procedures. It holds true for single queueing systems and for multiple-station networks, with open and closed flow of material.[3] It is valid for all processing time distributions, unlimited and limited buffer capacity and all queueing disciplines.[4]

Little's Law expresses the relationship between the mean work-in-process, the production rate, and the cycle time.[5] It represents a building block in several procedures, as for example in the mean value analysis. It states that the production rate, denoted by PR and measured in [quantity unit/time unit], equals the average number of workpieces in the system, denoted by L^S [quantity unit], divided by the cycle time of the complete system, T^S [time unit]:

$$PR = \frac{T^S}{L^S}.\tag{3.1}$$

Little's Law also holds in relation to the queue. It states that the production rate equals the mean number of workpieces in the queue, L^Q, divided by the time spent in the queue, T^Q:

---

[3] See Hopp and Spearman (2000, p. 223).
[4] See Bolch et al. (2006, p. 245).
[5] See Little (1961, pp. 383ff).

$$PR = \frac{T^Q}{L^Q}.$$ (3.2)

Arrival Theorem

The Arrival Theorem refers only to networks with exponential processing times and infinite buffer capacity. The Arrival Theorem is the short term for "Theorem of the distribution at arrival time". By "distribution" the distribution of the number of customers at a station is meant. The theorem was proposed and proven for closed product-form networks by Reiser and Lavenberg (1980) and Sevcik and Mitrani (1981).

The statement of the Arrival Theorem is that the probability mass function of the number of customers at a station—at the time-instant of the arrival of a customer—equals the probability mass function of the number of customers of the same network with one customer less in the system. More loosely speaking and only referring to the first moment, a customer arriving at a queue sees on average the same number of customers as an outsider would if the system had one less customer.[6]

Norton's Theorem

Norton's Theorem has its origins in electrical circuit theory and was transferred to queueing networks by Chandy, Herzog, and Woo (1975b). The concept is also known as the flow-equivalent server method, short-circuiting, and parametric analysis. The latter term corresponds to the title of their article.

According to Norton's Theorem, a composite node is created that includes all nodes except one. The parameters of that node are adapted so that this composite node mimics the pooled network. Then, as Norton's Theorem states, the reduced system has the same behavior as the original network. This way, the networks can be analyzed as two-station closed queueing networks. This theorem is useful because two-station systems are easier to analyze.

With Norton's Theorem, the exact analysis is possible under exponential processing times and unlimited buffer capacities.[7] Norton's Theorem is also used in many approximate approaches such as Marie's method which is again the basis for several other procedures.[8]

In the following, performance-analysis procedures are reviewed. Unless stated otherwise, the closed queueing networks are assumed to follow the first-come first-served queueing discipline, consist of single-server stations, have a single customer

---

[6]See Bolch et al. (2006, pp. 384f).

[7]See Bolch et al. (2006, pp. 410ff).

[8]Marie's method considers general processing times and works very well. For more details on this procedure, see Sect. 3.3.3.2 on page 35 and Marie (1979).

class, and in case of blocking, to follow the blocking after service discipline. On the first level, the literature review is divided into exact methods, see Sect. 3.2, and approximate methods, see Sect. 3.3. On the second level, the different settings are considered. Each section contains a table providing an overview of the procedures. Selected methods are described in more detail.

3.2 Exact Methods

Exact methods under the assumption of exponential processing times and unlimited buffer capacities are presented in Sect. 3.2.1. All other settings are considered in Sect. 3.2.2.

3.2.1 Product-Form Networks

Product-form networks are networks without capacity restrictions and exponential service times. The Gordon-Newell Theorem, the convolution algorithm and the mean value analysis are described in detail. Many other procedures are built upon these and are reviewed hereafter.

3.2.1.1 Gordon-Newell Theorem

The first considerations of the performance analysis of CQN go back to Gordon and Newell (1967). The Gordon-Newell Theorem[9] is based upon the Jackson Theorem proposed by Jackson (1957, 1963). The Jackson Theorem states a formula for the steady-state probabilities in open queueing networks with exponential processing times and infinite buffer space. His closed-form expression coined the term "product-form networks" as name for these networks, because the steady-state probabilities can be calculated by a product of terms. The notation of this section is given in Table 3.3.

Gordon and Newell (1967) transferred Jackson's results to closed queueing networks, with all other assumptions staying the same. The Gordon-Newell Theorem states a closed-form expression for the calculation of the steady-state probabilities for a given allocation of customers, see Eq. (3.3). The number of customers at station i is denoted by n_i.

$$P(n_1, n_2, \ldots, n_M) = \frac{1}{G(N)} \prod_{i=1}^{M} F_i(n_i). \tag{3.3}$$

---

[9] Although the Gordon-Newell Theorem is called a theorem, it allows the evaluation of the performance measures and is, therefore, sorted into this section.

Table 3.3 Notation

$\beta_i(n_i)$	Helping term for the calculation of $G(N)$
γ_i	Workload at station i, $\gamma_i = \frac{e_i}{\mu_i}$
$F_i(n_i)$	Weighting term for the calculation of $G(N)$
$G(N)$	Normalizing constant
e_i	Relative arrival frequency to station i, also-called visiting ratio of station i
μ_i	Processing rate at station i
n_i	Number of customers at station i
$P(n_1, \ldots, n_M)$	Probability of the customer allocation $n_1, \ldots, n_M$
s_i	Number of servers at station i

The term $G(N)$ is the so-called normalizing constant and $F_i(n_i)$ is a weighting term. $G(N)$ normalizes the sum of the probabilities of all states to 1. It is obtained by the sum of all weights that are applied to the steady-state probabilities, see Eq. (3.4).

$$G(N) = \sum_{\substack{\sum_{i=1}^{M} n_i = n}} \prod_{i=1}^{M} F_i(n_i) \tag{3.4}$$

The sum of all weights $\prod_{i=1}^{M} F_i(n_i)$ is received by considering all combinations of $(n_1, \ldots, n_M)$ that sum up to n. The term $F_i(n_i)$ is calculated by

$$F_i(n_i) = \frac{\left(\frac{e_i}{\mu_i}\right)^{n_i}}{\beta_i(n_i)}. \tag{3.5}$$

The ratio of the visiting ratio e_i and the processing rate μ_i, represents the workload, denoted by γ_i, i.e. $\gamma_i = \frac{e_i}{\mu_i}$. The function $\beta_i(n_i)$ is calculated by

$$\beta_i(n_i) = \begin{cases} n_i! & \text{if } n_i \leq s_i \text{ or } N \leq s_i, \\ s_i! \cdot s_i^{(n_i - s_i)} & \text{if } n_i > s_i, \end{cases} \tag{3.6}$$

where s_i denotes the number of servers at station i.[10] In the case of single servers ($\beta_i(n_i) = 1 \ \forall i, n_i$) and linear flow of material ($e_i = 1 \ \forall i$), the term $F_i(n_i)$ reduces to

$$F_i(n_i) = \left(\frac{1}{\mu_i}\right)^{n_i}. \tag{3.7}$$

---

[10]See Bolch et al. (2006, pp. 346f).

The results of Jackson (1963) and Gordon and Newell (1967) have been extended by Baskett, Chandy, Muntz, and Palacios (1975) to multiple customer classes, different queueing disciplines, and load-dependent exponential processing times, both for open and closed queueing networks.[11]

3.2.1.2   Convolution Algorithm

Based on the Gordon-Newell Theorem, Buzen (1973) developed the so-called convolution algorithm to calculate the normalizing constant, which is a part of the closed-form expression proposed by Gordon and Newell (1967). It needs less working memory and decreases the computation time.

Using Eqs. (3.4) and (3.7), the normalizing constant for CQN with single-server stations and a linear flow of material is given by

$$G(N) = \sum_{i=1}^{M} \prod_{n_i=1}^{N} \left(\frac{1}{\mu_i} \right)^{n_i}. \tag{3.8}$$

The pseudo code of the convolution algorithm for CQN with single-server stations and a linear flow of material is given in Algorithm 1.[12] The normalizing constant is initialized with $G(0) := 1$ and $G(n) := 0 \ \forall n$. $G(N)$ is computed by iterating over all stations i, and within i over all number of customers n.

Many computation steps are saved in comparison to calculating the normalizing constant according to Eq. (3.4), because several intermediate values can be omitted. Thus, the computation time is accelerated. Load-dependent service times can be applied as well, since each n is considered separately. This is used in several approximate techniques such as Marie's method.[13] The computation of the normalizing constant and the convolution algorithm is explained in detail in Tempelmeier and Kuhn (1993).[14]

Algorithm 1 Convolution algorithm of Buzen (1973) for single-server stations and linear flow of material

Initialization: $G(0) := 1; G(n) := 0 \quad \forall n$
for $i = 1$ **to** M **do**
    for $n = 1$ **to** N **do**
        $G(n) := \frac{1}{\mu_i} \cdot G(n-1) + G(n)$
    end for
end for

---

[11] See Baskett et al. (1975) and Bolch et al. (2006, pp. 353ff).

[12] See Tempelmeier and Kuhn (1993, p. 76).

[13] For a description of Marie's method, see page 35.

[14] See Tempelmeier and Kuhn (1993, p. 71ff).

Table 3.4 Notation

$L_i^S(n)$	Mean work-in-process at station i
$P_i(j_i\|n)$	Steady-state probability for j_i workpieces at station i when there are n in the system
$PR(n)$	Production rate with n workpieces in the system
T^S	Cycle time in the system
T_i^S	Cycle time at station i
T_i^{Stv}	Mean starving time at station i

Algorithm 2 Mean value analysis

Input: N, M, μ_i for $i = 1, \ldots, M$
Initialize $L_i^S(0) = 0 \ \forall i$
For $n = 1$ **to** N
 For $i = 1$ **to** M
  $T_i^S(n) = \frac{1}{\mu_i}\left[1 + L_i^S(n-1)\right]$
 Next i
 $PR(n) = \frac{n}{\sum_{i=1}^M T_i^S(n)}$
 For $i = 1$ **to** M
  $L_i^S(n) = PR(n) \cdot T_i^S(n)$
 Next i
Next n

3.2.1.3   Mean Value Analysis

Reiser and Lavenberg (1980) introduced the mean value analysis (MVA). It is a computationally very efficient method. In this procedure, the performance measures are calculated from mean values, hence the name. The notation used in this section is provided in Table 3.4.

This iterative procedure is founded on two theorems, the Arrival Theorem and Little's Law.[15] Based on the initializations and values of the former iteration, the mean cycle time, the production rate, and the mean work-in-process are calculated. The pseudo code of the MVA for one customer class, linear flow of material and single-server stations is given in Algorithm 2.[16]

The mean work-in-process is initialized by zero for no customers in the system, $L_i^S(0) = 0$. It is iterated over the number of customers, starting with $n = 1$ and it is incremented until the desired workpiece level N is reached.

First, the cycle time of station i, $T_i^S(n)$, is calculated for all stations. $T_i^S(n)$ results from the processing time multiplied by the mean number of customers at that station according to Little's Law, see Eq. (3.1). The mean number of customers is computed by $L_i^S(n-1) + 1$, which is the mean number of customers that an arriving customer sees plus the arriving customer itself. This is true according to the Arrival Theorem. The value of $L_i^S(n-1)$ is available from the former iteration.

---

[15]See Sect. 3.1.
[16]See Bolch et al. (2006, p. 388).

Next, the production rate is calculated, also corresponding to Little's Law. The cycle time of the system, $T^S(n)$, is obtained from the sum of the cycle time per station over all stations, $T^S(n) = \sum_{i=1}^{M} T_i^S(n)$. The mean work-in-process at station i, $L_i^S(n)$, is again calculated by Little's Law with the updated values of $PR(n)$ and $T_i^S(n)$.

The steady-state probabilities are also calculable from the mean value analysis. However, these are not needed for the calculation of the performance measures. The probability of j customers at station i, given that there are n customers in the system, is depicted in Eq. (3.9). These are obtained by rearrangement of the steady-state probabilities according to Gordon and Newell (1967),[17] see Eq. (3.9).

$$P(j_i|n) = \frac{PR(n)}{\mu_i} \cdot P(j_i - 1|n - 1) \tag{3.9}$$

The computation time of the mean value analysis rises with the number of stations and workpieces in the system. Further, it must be iterated over all $n < N$ to obtain the measures for the desired work-in-process level N.

Reiser (1979) extended the mean value analysis to multiple chains. Reiser (1981) considered queue-dependent servers. For this, he combined the mean value analysis with the convolution algorithm. It is called normalized convolution algorithm.

3.2.1.4 Further Methods

There are several more methods for CQN with exponential processing times and unlimited buffer space. The procedures are summarized in Table 3.5.

The parametric analysis by Chandy et al. (1975b)[18] is based on Norton's Theorem.[19] Building upon the parametric analysis, Akyildiz (1985) proposed a method needing less computation time. Chandy and Sauer (1980)[20] developed the so-called local balance algorithm for normalizing constants (LBANC). In this procedure, the normalizing constant is calculated iteratively. Based on this, Chandy and Sauer further proposed the so-called coalesce computation of normalizing constants (CCNC). It is designed to use less storage space.[21]

Yao and Buzacott (1985) considered CQN in which the parts follow a probabilistic shortest-queue scheme and derived the queue length. Lee and Seo (1998) computed the queue length distribution for two-station systems by the matrix geometric method. They modeled CQN as a stochastic event graph.

---

[17]See Bolch et al. (2006, p. 387).

[18]For a description of the method and implementation, see Bolch et al. (2006, pp. 410ff).

[19]See Sect. 3.1.

[20]See also Sauer and Chandy (1981).

[21]See also Bolch et al. (2006, pp. 369f).

Table 3.5 Exact methods considering exponential processing times and unlimited buffer space

Authors	C	M	S	D	BS	BM	R	BP	N	OA
Akyildiz (1985)	CR	M	S	E	U	–	ex	PA	EPA	
Chandy et al. (1975b)	CR	M	S	E	U	–	ex	NT	PA	
Chandy and Sauer (1980)	CR	M	S	E	U	–	ex	MVA	LBANC, CCNC	[a]
Boxma, Kelly, and Konheim (1984)	CL	M	S	E	U	–	ex	LST		[b]
Buzen (1973)	CR	M	S	E	U	–	ex	GNT	C	
Chow (1980)	CL	2	S	E	U	–	ex	LST		
Gordon and Newell (1967)	CR	M	S	E	U	–	ex		GNT	
Lee and Seo (1998)	CL	2	S	E	U	–	ex	MC		[c]
Reiser (1979)	CR	M	S	E	U	–	ex		MVA	[d]
Reiser (1981)	CR	M	S	E	U	–	ex	MVA,C	NCA	[e]
Reiser and Lavenberg (1980)	CR	M	S	E	U	–	ex		MVA	[a]
Schassberger and Daduna (1983)	CL	M	S	E	U	–	ex	LST		
Yao and Buzacott (1985)	CF	M	S	E	U	–	ex	C		[a,f]

Abbreviations:

C Configuration (*CF* flexible manufacturing system, *CL* closed and linear, *CR* closed and routing)
M number of stations (*M* arbitrary many), *S* number of servers (*S* single), *D* Distribution of the processing times (*E* exponential), *BS* Buffer space (*U* unlimited), *BM* Blocking mechanism, *R* Result (*ex* exact), *BP* base procedure (*C* convolution algorithm, *CCNC* coalesce computation of normalizing constants, *EPA* extended parametric analysis, *GNT* Gordon-Newell Theorem, *LBANC* local balance algorithm for normalizing constants, *LST* Laplace Stieltjes transformation, *MC* Markov chains, *MVA* mean value analysis, *NCA* normalized convolution algorithm, *NT* Norton's Theorem, *PA* parametric analysis), *N* name of procedure, *OA* other aspects, see below:

[a] Multiple-class customers
[b] Calculation of the cycle time distribution
[c] Queue-length distribution
[d] Window flow control; multiple chains
[e] Queue-dependent servers
[f] Probabilistic shortest queue, service discipline: random

A completely different approach to the performance analysis is the calculation of the exact cycle time distribution via Laplace transformations. The concept was introduced by Chow (1980) for two-station systems. Chow found that the cycle time distribution has Erlangian distribution when the processing times are exponential and buffers are infinite. Schassberger and Daduna (1983) and Boxma et al. (1984) extended this approach to more than two stations.

3.2.2   Non-Product-Form Networks

Non-product-form networks are networks which do not fulfill the assumptions of product-form networks. Exact methods exist for CQN with exponential processing times and finite buffer capacities, as well as for phase-type distributed processing

Table 3.6 Exact methods for non-product-form networks

Authors	C	M	S	D	BS	BM	R	BP	N	OA
Akyildiz (1987)	CL	2	M	E	L	BAS	ex	MSS		
Akyildiz (1988c)	CR	2	S	E	L	BAS	ex		MSS	
Akyildiz and Huang (1993)	CR	M	S	E	L	BAS	ex	NC, MSS		[a]
Akyildiz and von Brand (1994)	CL	2	M	E	L	BAS	ex	GBE		[b]
Altiok (1996)	OL	2	1	C	L	BAS	ex	MC		
Balsamo and Clò (1998)	CR	M	S	E	L	BBS, BAS, RS	ex	PFS		
Balsamo and Donatiello (1989)	CL	2	S	E	L	BBS	ex			[c]
Boxma (1983)	CL	2	S	1st G, 2nd E	U	–	ex			[c]
Boxma and Donk (1982)	CL	2	S	E	U	–	ex	LST		
Carbini, Donatiello, and Iazeolla (1986)	CL	2	S	1st E, 2nd C	U	–	ex	LST		
Clò (1998)	CR	M	S	E	L	RS	ex	MVA		
Daduna (1984)	CL	2	S	1st G, 2nd E	U	–	ex	LST		
Daduna (1986)	CL	2	S	1st ME 2nd G	U	–	ex	LST		
Kleinrock (1975)	CR	M	S	PH	U	–	ex	MC		
Zhuang, Buzacott, and Liu (1994)	CL	2	S	E	L	BAS	ex	MVA		

Abbreviations:

C Configuration (*CL* closed and linear, *CR* closed and routing, *OL* open linear), *M* number of stations (*M* arbitrary many), *S* number of servers (*M* multiple, *S* single), *D* Distribution of the processing times (*C* Coxian, *E* exponential, *G* general, *ME* mixed Erlang, *PH* phase-type), *BS* Buffer space (*L* limited, *U* unlimited), *BM* Blocking mechanism (*BAS* blocking after service, *BBS* blocking before service, *RS* repetitive service blocking), *R* Result (*ex* exact), *BP* base procedure (*LST* Laplace Stieltjes transformation, *MC* Markov chains, *MSS* Matching state space, *NC* normalizing constant), *N* name of procedure (*MSS* matching state space), *OA* other aspects, see below:

[a] Service discipline: processor sharing or infinite servers, multiple customer classes
[b] Symmetric scheduling
[c] Recursive algorithm

times and infinite buffer capacities. Most exact procedures for non-product-form networks presuppose two-station systems. Table 3.6 provides an overview.

There are several exact methods for CQN with exponential processing times and finite buffers with two or more stations. Akyildiz and Huang (1993) assumed several types of service disciplines. They obtained exact product-form solutions for the equilibrium-state probabilities by matching the state space. Balsamo and Clò (1998) proposed a convolution algorithm for various types of blocking mechanisms. Clò (1998) developed an algorithm based on the mean value analysis for closed queueing networks with repetitive service blocking. Equivalence relations and properties were derived to calculate the performance measures.

There are further methods that are restricted to two-station systems. Akyildiz (1987) converted the state space of a CQN with finite buffer spaces and exponential processing times into an equivalent product-form network. Akyildiz (1988c) accomplished the exact analysis by matching the state space. In the approach of Akyildiz and von Brand (1994), the global balance equations are used for CQN with symmetric scheduling. Balsamo and Donatiello (1989) calculated the cycle time for two-station systems with a blocking-before-service mechanism. Zhuang et al. (1994) built upon the mean value analysis. They considered blocking by adapting the steady-state probabilities, calculated the mean queue lengths from these, and applied these expressions to the mean cycle time in the MVA.

The analysis of two-station systems by Laplace-Stieltjes transforms has been extended to one general server by Boxma and Donk (1982), Boxma (1983) and Daduna (1984). Carbini et al. (1986) assumed that the service time of one server follows a Coxian distribution. Daduna (1986) assumed the first server to have a mixed-erlang distribution and the second to follow a general distribution.

Altiok (1996) showed how to analyze open two-station subsystems with finite buffer and Cox-2 distributed processing times by means of Markov chains. Kleinrock (1975) introduced the only existing exact method for CQN with more than two stations and phase-type distributed processing times, also by Markov chains. However, there is no exact method for CQN with general processing times and finite buffer space. We will build upon the Markov-chain approaches of Kleinrock (1975) and Altiok (1996) in order to analyze CQN with phase-type distributions and finite buffer capacities, see Chap. 5.

3.3   Approximate Methods

Many approximate methods have been developed for closed queueing networks with exponential and general processing times, as well as with limited and unlimited buffer capacities. Each of the four settings is considered separately in the following.

3.3.1   Exponential Distribution and Infinite Buffer Capacity

Approximate methods for CQN with exponential distribution and infinite buffer capacity are faster than exact methods. Furthermore, they represent the basis for several more general procedures. Table 3.7 provides an overview of these methods.

Koenigsberg and Lam (1976) applied CQN to loading and discharging operations of gas vessels in ports. They used the normalizing constant for the analysis. Furthermore, they considered non-exponential distributions in simulation in order to examine the effect of the coefficient of variation on the performance measures.

Several approximate methods are based on mean value analysis. Bard (1979) and Schweitzer (1979) developed approximation formulas for the cycle time in multi-class closed queueing networks. This approach was improved by Bard (1981).

Table 3.7 Approximate methods considering exponential processing times and unlimited buffer space

Authors	C	M	S	D	BS	BM	R	BP	N	OA
Akyildiz and Bolch (1988)	CR	M	M	E	U	–	app	MVA		[a]
Bard (1979)	CR	M	S	E	U	–	app	MVA		[b,c]
Bard (1981)	CR	M	S	E	U	–	app	MVA		
Bolch, Fleischmann, and Schreppel (1987)	CR	M	S	E	U	–	app	SUM		
Bolch and Fischer (1993)	CR	M	M	E	U	–	app		SUM	
Chen-Hong (1999)	CR	M	S	E	U	–	app			[d]
Duenyas and Hopp (1990)	CL	M	S	E	U	–	app			[e]
Koenigsberg and Lam (1976)	CL	M	M	E	U	–	app	NC		
Schmidt (1997)	CR	M	M	E	U	–	app	MVA, NT		[b,f]
Schweitzer (1979)	CR	M	S	E	U	–	app	MVA		
Schweitzer, Seidmann, and Shalev-Oren (1986)	CR	M	S	E	U	–	app	MVA		[b]
Spearman (1991)	CL	M	S	E	U	–	app	MVA		[g]
Strelen (1989)	CL	M	M	E	U	–	app	MVA		[h]
Suri and Desiraju (1997)	CF	M	S	E	U	–	app	MVA		[i]
Suri and Hildebrant (1984)	CF	M	M	E	U	–	app	MVA	MVAQ	

Abbreviations:

C Configuration (*CF* flexible manufacturing system, *CL* closed and linear, *CR* closed and routing), *M* number of stations (*M* arbitrary many), *S* number of servers (*M* multiple, *S* single), *D* Distribution of the processing times (*E* exponential, *LE* load-dependent exponential), *BS* Buffer space (*U* unlimited), *BM* Blocking mechanism (*BAS* blocking after service, *BBS* blocking before service), *R* Result (*app* approximate) *BP* base procedure (*C* convolution algorithm, *D* decomposition, *MM* Marie's method, *MVA* mean value analysis, *NC* normalizing constant, *PA* parametric analysis, *SUM* summation method), *N* name of procedure (*MVAQ* Mean Value Analysis of Queues, *RECAL* Recursion by Chain Algorithm, *SCAT* Self Correcting Approximation Technique, *OA* other aspects, see below:

[a] Multiple servers
[b] Multiple-class customers
[c] Priority, processor sharing
[d] Variance of cycle time
[e] Variance of output
[f] Class-dependent server
[g] Bounds for performance measures
[h] Higher moment of queue length and waiting time
[i] State-dependent routing, infinite server

Schweitzer et al. (1986) revealed a correction term for the approximate mean value analysis by Schweitzer (1979). Neuse and Chandy (1981) and Chandy and Neuse (1982) developed the so-called Self Correcting Approximation Technique (SCAT) further improving the Bard-Schweitzer Approximation.[22] Furthermore, Chandy and Neuse considered general distributions for the processing times.[23]

---

[22] See also Bolch et al. (2006, pp. 422, 427).

[23] For this, see Sect. 3.3.3.4.

Suri and Hildebrant (1984) presented an approximation for flexible manufacturing systems, called mean value analysis of queues (MVAQ). Strelen (1989) extended the MVA to the calculation of higher-order moments regarding the queue length and waiting time for linear closed queueing systems. Akyildiz and Bolch (1988) based their approximation for CQN with multiple servers on the estimation of the conditional marginal probabilities.

Spearman (1991) developed the so-called congestion model for linear CQN with processing times that are either exponentially distributed or that follow any other distribution with an increasing failure rate.[24] He provided lower and upper bounds for the mean cycle time and the throughput. Thus, he created best-case and worst-case scenarios. Schmidt (1997) proposed a procedure for exponential, class-dependent multiple servers. He approximated the cycle time and used it in an MVA-based procedure. Suri and Desiraju (1997) developed an iterative method based on the mean value analysis for flexible manufacturing systems with state-dependent routing and infinite servers.

Bolch et al. (1987) proposed an approximation for the mean work-in-process, which was then used in the mean value analysis as well.[25] It allows the omission of iterations over those workpiece levels that are not of interest. It is called the summation method (SUM). Bolch and Fischer (1993) extended the SUM by a bottleneck analysis. The procedure is called Bottapprox. In this procedure, the number of iterations is further reduced by initializing λ with the processing rate of the bottleneck station. Bolch et al. (2006) showed that Bottapprox is faster, but slightly worse than SUM. SUM, Bottapprox and SCAT were later extended to analyze CQN with general processing times (ESUM, EBOTT, and ESCAT), see Sect. 3.3.3.

Duenyas and Hopp (1990) estimated the variance of the production rate with Laplace transforms. Chen-Hong (1999) approximated the mean and the variance of the cycle time using a Bayesian statistical approach.[26]

3.3.2   *Exponential Distribution and Finite Buffer Capacity*

In the following, methods for CQN with exponential processing times and finite buffers are presented. The challenge in this case consists of calculating the blocking effect. Two widely used methods to overcome this problem are decomposition and mean value analysis. The extended mean value analysis by Yüzükirmizi (2005), as one example, is explained in Sect. 3.3.2.1. The decomposition technique is investigated in Chap. 4. In Sect. 3.3.2.2, methods incorporating the aforementioned assumptions are summarized.

---

[24]For more information on increasing failure rates, see Marshall and Olkin (2007).

[25]See also Bolch et al. (2006, pp. 440ff).

[26]See also Chen, George, and Tardif (2001).

Table 3.8 Notation

b_i	Buffer capacity at station i
λ_n	Production rate with n customers in the system
$\mu_i^e(n)$	Effective processing rate with n customers in the system
$\mu_i(n)$	Processing rate with n customers in the system
n_i	Number of customers at station i
$P_i^B(n)$	Blocking probability with n customers in the system
$P_i(n_i \mid n)$	Steady-state probability of n_i customers at station i when there are n customers in the complete system
s_i	Number of servers at station i
$T_i^S(n)$	Cycle time at station i with n customers in the system

3.3.2.1 Extended Mean Value Analysis

Yüzükirmizi (2005) used the mean value analysis to tackle the problem of blocking. His idea was to approximate the blocking probability, to estimate the mean blocking time from this, and to use the latter to estimate the cycle time. The notation of this section is given in Table 3.8.

The algorithm that Yzkirmizi proposed is summarized in Algorithm 3. The effective service rates $\mu_i^e(n)$ are initialized with the given service rates, $\mu_i^e(n) = \mu_i$ for $n = 1$. The probability that the buffer at station i, b_i, is full and the server is busy with $n = 0$ parts in the system, equals $P_i^B(0) = 0 \ \forall i$. In further iterations, $n \geq 2$, the blocking time is included as part of the effective processing time of the downstream station. For this, the blocking probability $P_{i+1}^b(n - 1)$ of the previous iteration is used. The conditional probability of having j parts at station i, when there are n parts in the system, is initialized with $P_i(j \mid n) = 0 \ \forall i, j = 0, n = 0$.

In the extended MVA, the effective processing rate is used instead of the processing rate, see Eq. (3.10). It is given by the reciprocal of the processing and blocking time. The mean blocking time equals the probability that station i is blocked, multiplied by the residual processing time (which equals the mean processing time[27]), $P_{i+1}^B(n - 1) \cdot \frac{1}{\mu_{i+1}(n)}$.

$$\frac{1}{\mu_i^e(n)} = \begin{cases} \frac{1}{\mu_i} & \text{if } n = 1, \\ \frac{1}{\mu_i(n)} + P_{i+1}^B(n - 1)\frac{1}{\mu_{i+1}(n)} & \text{if } 2 \leq n \leq N. \end{cases} \tag{3.10}$$

The effective processing rate is employed to calculate the mean cycle time as shown in Eq. (3.11). Using Little's Law, it is calculated by the relation of the work-in-process j to the estimate of the production rate subject to j for all possible j. Therefore, the fraction is weighted by the estimated probability for the considered work-in-process j, $P_i(j_i - 1 \mid n - 1)$.

---

[27] See the Markov property on page 92.

Algorithm 3 Extended mean value analysis by Yüzükirmizi (2005)

1: **procedure** EMVA($M, \mu_i \forall i, K_i, c_i$)
2:      $P_i(0|0) = 1, P_i^B(0) = 0$
3:      **for** $i = 1$ to n **do**
4:          **for** $i = 1$ to M **do**
5:              $\mu_i(n) = \begin{cases} n\mu_i & \text{if } n \leq s_i, \\ s_i\mu_i & \text{if } n > s_i. \end{cases}$
6:          **end for**
7:          **for** $i = 1$ to M **do**
8:              $\frac{1}{\mu_i^e(n)} = \begin{cases} \frac{1}{\mu_i} & \forall i \wedge n = 1, \\ \frac{1}{\mu_i(n)} + P_h^B(n-1)\frac{1}{\mu_h(n)} & \forall i \wedge 2 \leq n \leq N. \end{cases}$
9:
10:                          with $h = \begin{cases} i + 1 & \text{for } i \leq M, \\ 1 & \text{for } i = M \end{cases}$
11:          **end for**
12:          **for** $i = 1$ to M **do**
13:              $T_i^S(n) = \sum_{j=1}^{n} \frac{j}{\mu_i^e(j)} P_i(j-1|n-1)$
14:          **end for**
15:          **for** $i = 1$ to M **do**
16:              $\lambda_n = \frac{n}{\sum_{i=1}^{M} T_i^S(n)}$
17:          **end for**
18:          **for** $i = 1$ to M **do**
19:              $P_i(j|n) = \frac{\lambda(n)}{\mu_i^e(j)} P_i(j-1|n-1)$ for $j = 1, \ldots, n$
20:          **end for**
21:          $P_i(0|n) = 1 - \sum_{j=1}^{n} P_i(j|n)$
22:
23:          **for** $i = 1$ to M **do**
24:              $P_i^B(n) = 1 - \sum_{j=0}^{b_i+s_i} P_i(j|n)$
25:          **end for**
26:          **return** λ_n
27:      **end for**
28: **end procedure**

$$T_i^S(n) = \sum_{j=1}^{n} \frac{j}{\mu_i^e(j)} P_i(j_i - 1|n-1) \qquad (3.11)$$

The steady-state probabilities for $j_i = 1, \ldots, b_i + s_i$, denoted by $P_i(j_i|n)$, are calculated according to Eq. (3.9). Since the number of workpieces in the system n may be higher than the station capacity $b_i + s_i$, i.e. $n > b_i + s_i$, the steady-state probabilities for the states from $j_i = 1, \ldots, b_i + s_i$ do not add up to 1. Yüzükirmizi (2005) states that the blocking probability is estimated by the surplus probability mass of those customers that would not fit into the station because of the given station capacity, see Eq. (3.12).

$$P_i^B(n) = 1 - \sum_{j_i=0}^{b_i + s_i} P_i(j_i|n) \qquad (3.12)$$

3.3.2.2 Further Methods

Table 3.9 provides an overview of approximate methods considering exponential processing times and limited buffer space. Akyildiz (1988c) transferred the finite CQN into an equivalent infinite CQN by transforming the state space. The method is exact for two-station systems. Akyildiz (1989) presented an approximate product-form solution for the equilibrium-state probabilities in CQN with BAS. He obtained the solution by normalizing the infeasible states that violate the station capacities.

Akyildiz and Liebeherr (1989) extended Norton's Theorem to blocking. In this approach, each subnetwork is replaced by a composite node with infinite capacity and adapted parameters of the processing time distribution. Suri and Diehl (1986) also based their analysis on the flow-equivalent-server concept and further proposed a variable buffer size model. The procedures by Onvural and Perros (1989a) and Perros et al. (1988) are founded on Norton's Theorem as well. The method by Onvural and Perros can handle both BBS and BAS. In order to take the blocking effect into account, they set up equations that were solved as a fixed-point problem.

Yao and Buzacott (1986b) adapted the normalizing constant in order to consider finite buffers. Perros et al. (1988) derived an approximation of the queue-length probability distribution. On that basis, the parameters were adapted. Tempelmeier et al. (1989) developed a correction term for the queue length distribution resulting from the Gordon-Newell Theorem to account for finite buffer capacities in flexible manufacturing systems.

The concept of decomposition was first applied to CQN by Dallery and Frein. Frein and Dallery (1989) proposed a decomposition approach for CQN with BBS. Dallery and Frein (1989) extended that approach to load-dependent exponential processing times. Bouhchouch et al. (1993) developed a decomposition approach considering BAS. In these approaches, a bisection search was carried out to find an arrival rate to an open queueing system, which is analyzed instead of the CQN. Zhuang and Hindi (1993) and Liu et al. (1993) also proposed a decomposition method under blocking-after-service for flexible manufacturing systems. The subsystems of the aforementioned approaches were analyzed by a M/M/1/K model. In the decomposition approach by Tolio and Gershwin (1998), each finite-buffer subsystem was analyzed by setting up and solving Markov chains.

Akyildiz (1988b) approximated the performance measures by incorporating the blocking time into the mean cycle time. The mean blocking time was estimated by the residual service time. Zhuang and Hindi (1990, 1991) considered multi-class customers, multiple servers, job reversible routing, blocking-and-recirculate and blocking-after-service mechanisms in flexible manufacturing systems. These were evaluated by an approximate mean value analysis. They adapted the mean cycle time by incorporating the residual service time at the blocking station, and

Table 3.9 Approximate methods considering exponential processing times and limited buffer space

Authors	C	M	S	D	BS	BM	R	BP	N	OA
Akyildiz (1988b)	CR	M	S	E	L	BAS	app	MVA		
Akyildiz (1988c)	CR	M	S	E	L	BAS	app		MSS	a
Akyildiz (1989)	CR	M	S	E	L	BAS	app	PFS		
Akyildiz and Liebeherr (1989)	CR	M	S	E	L	BAS	app	NT		
Bouhchouch, Frein, and Dallery (1993)	CL	M	S	E	L	BAS	app	D		
Dallery and Frein (1989)	CL	M	S	LE	L	BAS	app	D		b
Frein and Dallery (1989)	CL	M	S	E	L	BBS	app	D		
Gonzales (1997)	CR	M	S	E	L	BAS	app	D		
Liu, Zhuang, and Buzacott (1993)	CL	M	S	E	L	BAS	app	NT,D		b
Onvural and Perros (1989a)	CR	M	S	E	L	BAS, BBS	app	NT		
Osorio and Bierlaire (2009)	CR	M	S	E	L	BAS	app	D		
Perros, Nilsson, and Liu (1988)	CR	M	S	E	L	BAS	app	NT		
Suri and Diehl (1986)	CL	M	S	E	L	BAS	app	NT		b
Tempelmeier, Kuhn, and Tetzlaff (1989)	CF	M	M	E	L	BAS	app	NC		
Tolio and Gershwin (1998)	CL	M	S	E	L	BAS	app	MC,D		
Yao and Buzacott (1986b)	CF	M	S	E	L	BBS	app	NC		
Yüzükirmizi (2006)	CR	M	M	E	L	BAS	app	MVA		
Zhuang and Hindi (1990)	CF	M	M	E	L	BR	app	MVA		c,d
Zhuang and Hindi (1991)	CF	M	M	E	L	BAS	app	MVA		c,e
Zhuang and Hindi (1993)	CF	M	M	E	L	BAS	app	D		c,e
Zhuang et al. (1994)	CL	M	S	E	L	BAS	app	MVA		f

Abbreviations:

C Configuration (*CF* flexible manufacturing system, *CL* closed and linear, *CR* closed and routing), *M* number of stations (*M* arbitrary many), *S* number of servers (*M* multiple, *S* single), *D* Distribution of the processing times (*E* exponential, *LE* load-dependent exponential), *BS* Buffer space (*L* limited), *BM* Blocking mechanism (*BAS* blocking after service, *BBS* blocking before service, *BR* block-and-recirculate), *R* Result (*app* approximate), *BP* base procedure (*D* decomposition, *EM* Expansion Method, *MC* Markov chains, *MVA* Mean value analysis, *NC* normalizing constant, *NT* Norton's Theorem, *PFS* product-form solution), *N* name of procedure (*D* Decomposition, *MSS* Matching State Space), *OA* other aspects, see below:

[a]Exact for two stations
[b]One buffer infinite
[c]Multi-class customers
[d]Job reversible routing
[e]State-dependent routing
[f]Exact for two-station systems with blocking

thus, accounted for the blocking time. Zhuang et al. (1994) considered BAS in an approximate MVA. In their approximation, the blocking effect of the cyclic queue is taken into account by adapting the steady-state probabilities, calculating the mean queue lengths from these, and using this expression for the mean cycle time in the MVA.

Gonzales (1997) approximated the performance measures by expanding and decomposing the network. The method is applicable to different topologies with

Table 3.10 Notation

c_B^2	Coefficient of variation of the processing time distribution
c_A^2	Coefficient of variation of the inter-arrival time distribution
δ	Helping term
e_i	Visiting ratio of station i
$L^S(n)$	Estimated mean work-in-process
p_{ij}	Routing probability from station i to station j
$\pi(n_1, \ldots, n_M)$	Marginal probability for the distribution of customers over the stations $(n_1, \ldots, n_M)$
$P(n)$	Steady-state probability for n workpieces
ρ	Utilization

single-server stations. Osorio and Bierlaire (2009) analyzed CQN of arbitrary topology with multiple-server queues in a congestion model.

3.3.3 General Distribution and Infinite Buffer Capacity

In this section, approximate procedures for CQN with general processing times and infinite buffers are presented. In this setting, blocking does not play a role but, therefore, arbitrary variabilities of the processing times make the performance analysis more challenging.

In the following, the diffusion approximation, Marie's method, and the modified MVA are described in detail. Further, an overview of further procedures is given. The diffusion approximation is often used in queueing systems and yields approximations for the steady-state probabilities. Marie's method provides very good approximations for $c^2 > 0.5$ and has been applied in several other procedures. The modified MVA serves as an example for various approaches using the MVA for CQN with general service times.

3.3.3.1 Diffusion Approximation

The diffusion approximation provides expressions for the steady-state probabilities of a queue with general arrival and service times and infinite waiting space. The notation is presented in Table 3.10.

The diffusion approximation relies on the central limit theorem. This theorem asserts that the sum of independent random variables can be approximated by a normal distribution. This is regardless of the particular distribution of the random variables. More specifically, this means that n independent and identically distributed random variables X_i $(i = 1, \ldots, n)$ with mean $E[X_i] = \mu$ and variance $Var[X_i] = \sigma$, where n is a sufficiently large number, are approximately normally distributed with mean $n\mu$ and variance $n\sigma$.[28] Based on this assumption, Kobayashi (1974) and Reiser and Kobayashi (1974) derived the density function of the stochastic process of a queue, that was modeled as a continuous process. The discretization of the density function yields the equations for the probabilities, see Eqs. (3.13)–(3.16).

$$P(n) = \begin{cases} 1 - \rho & \text{if } n = 0, \\ \rho(1 - \delta)\delta^{n-1} & \text{if } n > 0 \end{cases} \tag{3.13}$$

with

$$\rho = \frac{-2(1 - \rho)}{c_B^2 + \rho c_A^2} \tag{3.14}$$

as estimate for the utilization, further with

$$\delta = \exp(\rho) \tag{3.15}$$

and finally with

$$c_A^2 = 1 + \sum_{i=1}^{M} \left(c_B^2(i) - 1\right) \cdot p_{ij}^2 \cdot e_j \cdot e_i^{-1} \tag{3.16}$$

as estimate for the coefficient of variation of the inter-arrival time. This expression can, moreover, be used in a queueing model. Thus, the application of the diffusion approximation for the analysis of subsystems in a decomposition approach is enabled.

For a high utilization, Eq. (3.13) does not provide good results. Based on the work by Gelenbe (1975) and Mitzlaff (1997) derived more accurate formulas by imposing different boundary conditions for the approximately normally distributed continuous-state process, see Eq. (3.17). The parameters of (3.17) equal those of Eq. (3.13), which means, these are given in Eqs. (3.14)–(3.16).

$$P(n) = \begin{cases} 1 - \rho & \text{if } n = 0, \\ \rho\left[1 - \dfrac{1}{\gamma}(\delta - 1)\right] & \text{if } n = 1, \\ -\dfrac{\rho}{\gamma\delta^2}(1 - \delta)^2\delta^n & \text{if } n > 1 \end{cases} \tag{3.17}$$

Bolch et al. (2006) applied the diffusion approximation to CQN.[29] The procedure is carried out as follows:

---

[28] See Stewart (2009, p. 184).
[29] See Bolch et al. (2006, pp. 463ff).

Table 3.11 Notation

$\lambda_i(n_i)$	Load-dependent arrival rate to station i when there are n_i customers at station i
$\nu_i(n_i)$	Load-dependent departure rate of station i when there are n_i customers at station i
$\lambda_c^{(i)}(n - n_i)$	Load-dependent arrival rate of $n - n_i$ customers at the composite node c

1. The utilization at each node is approximated by a product-form algorithm (neglecting the coefficient of variation).
2. The marginal probabilities $P_i(n_i)$ are estimated according to the diffusion approximation in Eq. (3.17) [or (3.13)] with ρ from step 1.
3. Computation of the steady-state probabilities

$$\pi(n_1, \ldots, n_M) = \prod_{i=1}^{M} P(n_i)\frac{1}{G} \tag{3.18}$$

with the condition that

$$G = \sum_{\substack{M \\ \sum_{i=1}^{M} n_i = n}} \pi(n_1, \ldots, n_M) = 1. \tag{3.19}$$

4. The performance measures are computed from the steady-state probabilities obtained in step 2.

By applying the steady-state-probabilities estimate of Eq. (3.17), we receive the following approximation for the mean work-in-process:

$$L^S(n) = \sum_{j=1}^{n} P(j) \cdot j = \rho\left[1 + \frac{\rho c_A^2 + c_B^2}{2(1-\rho)}\right]. \tag{3.20}$$

In this method, all states of the system, i.e. all workpiece allocations, have to be derived in order to calculate the performance measures.

3.3.3.2 Marie's Method

Marie (1979, 1980) developed a very precise approximation method for the performance evaluation of general closed queueing networks. The notation is given in Table 3.11.

In this procedure, general processing times are converted into load-dependent exponential processing times. It is distinguished between arrival and departure rates, denoted by $\lambda_i(n_i)$ and $\nu_i(n_i)$ respectively. Further, for each node i, the rest of the network is aggregated to one composite node according to Norton's Theorem. Then, a product-form algorithm for load-dependent service times is applied as, for example, the convolution algorithm.

The fact that the customers circulate in the system is used for the following statement: When n_i customers reside at node i, there must be $(n - n_i)$ customers at the composite node c. Assuming load-dependent arrival rates for $\lambda_i(n_i)$ and requiring conservation of flow, as a consequence, Eq. (3.21) holds true:

$$\lambda_i(n_i) = \lambda_c^{(i)}(n - n_i) \tag{3.21}$$

Equation (3.21) states that the arrival rate of node i with n_i customers, $\lambda_i(n_i)$, equals the arrival rate of the remaining $n - n_i$ customers at the composite node c, denoted by $\lambda_c^{(i)}(n - n_i)$. The system is then modeled as a birth-death process. The birth rate corresponds to the arrival rate λ_i and the death rate represents the departure rate ν_i, see Marie and Stewart (1977). Under these assumptions it holds that

$$\nu_i(n_i) \cdot P(n_i) = \lambda_i(n_i - 1) \cdot P(n_i - 1). \tag{3.22}$$

For the birth-death process, closed-form solutions exist, see Eqs. (3.23) and (3.24).

$$P(n_i) = P(0) \cdot \prod_{j=0}^{n-1} \frac{\lambda_i(j)}{\nu_i(j+1)} \tag{3.23}$$

$$P(0) = \frac{1}{1 + \sum_{j=1}^{n} \prod_{k=0}^{j-1} \frac{\lambda_i(k)}{\nu_i(k+1)}} \tag{3.24}$$

Marie's method is an iterative procedure. First, the load-dependent service rates are initialized by the predefined processing rates (neglecting the coefficient of variation). Subsequently, a product-form procedure (e.g. the convolution algorithm) is applied to calculate the arrival rates λ_i. The λ_i are then employed to compute the ν_i according to the specified service time distribution. The ν_i serve as input for the product-form algorithm in order to update the λ_i. In each iteration, the corresponding number of customers is estimated. The adaption of λ_i and ν_i is repeated until the estimated mean number of customers in the network deviates only by a suitable tolerance ϵ from the specified number of customers.[30] The method is computationally efficient and yields only small deviations from the simulation for $c^2 \geq 0.5$.

---

[30]Bolch et al. (2006, p. 492).

Table 3.12 Notation

$c[T_i]$	Coefficient of variation of the processing time
e_i	Visiting ratio of station i
$E[T_i]$	Expected value of the processing time at station i
L_i^Q	Mean number of jobs in the queue of station i
L_i^S	Mean number of jobs at station i
T_i	Random variable describing the processing time distribution at station i
T_i^S	Mean cycle time at station i with n jobs in the system
$\rho_i(n)$	Utilization at station i with n jobs in the system
$Var[T_i]$	Variance of the processing time

3.3.3.3 Modified MVA

Curry and Feldman (2008) adapted the mean value analysis in order to analyze networks with general processing times. They proposed an estimation for the mean cycle time in CQN with general service times. The notation is given in Table 3.12.

The mean cycle time is composed of

- The remaining service time of the current job in service:
  It equals the probability that the server is busy, estimated by $\rho_i(n-1)$, multiplied by the estimated mean residual life time,[31] given by $\frac{E[T_i^2]}{2E[T_i]}$.
- The complete service time of all jobs in the queue:
  It is calculated from the mean service time, $E[T_i]$, multiplied by the mean number of jobs in the queue, that is $L_i^Q \approx L_i^S(n-1) - \rho_i(n-1)$.[32]
- And the expected service time of the arriving job, $E[T_i]$.

The mean cycle time as the sum of the aforementioned components is given in Eq. (3.25).

$$T_i^S(n) = \rho_i(n-1) \cdot \frac{E[T_i^2]}{2E[T_i]} + E[T_i] \cdot \left(L_i^S(n-1) - \rho_i(n-1)\right) + E[T_i] \quad (3.25)$$

with

$$\rho_i(n) = \frac{n \cdot e_i \cdot E[T_i]}{\sum_{j=1}^{n} e_j \cdot T_j^S(n)} \quad (3.26)$$

and

$$E[T_i^2] = E[T_i]^2 + Var[T_i] = E[T_i]^2 + (c^2[T_i] \cdot E[T_i]^2) = E[T_i]^2 \cdot (c^2[T_i] + 1) \quad (3.27)$$

---

[31] See Altiok (1996, p. 30).

[32] Here, $\rho_i(n-1)$ is the probability that the server is occupied. As no blocking occurs, the probability that a job receives service equals the probability that the server is busy. In case of blocking, the utilization and the blocking probability would have to be subtracted from the mean work-in-process.

Plugging (3.27) and (3.26) into (3.25), the resulting expression for the mean cycle time depends only on the given values, see Eq. (3.28).

$$T_i^S(n) = E[T_i] + \frac{(n-1) \cdot e_i}{\sum\limits_{j=1}^{n} e_j \cdot T_j^S(n-1)} \cdot$$

$$\left[E[T_i] \cdot T_i^S(n-1) + \frac{E[T_i]^2 \cdot (c^2[T_i]+1)}{2} \right] \tag{3.28}$$

With the reformulation of the cycle time given in Eq. (3.28), the mean value analysis is carried out. The procedure has also been extended to multi-class networks with general processing times.[33]

3.3.3.4   Further Methods

In Table 3.13, an overview of the methods which consider general processing times and infinite buffer space is given.

Gaver and Shedler (1973b) estimated the performance analysis of two-station systems by a diffusion approximation with a focus upon processor-utilization estimation. The approach has a deficiency in the estimation if $c^2 > 1$. Therefore, Gaver and Shedler (1973a) altered the approximation such that it is more accurate. They considered hyper-exponentially distributed service times. In the approximation by Gelenbe (1975), additional equations were introduced to represent the behavior of empty queues. This reduced the dependency of the model on heavy traffic assumptions.

The approach by Chandy et al. (1975a) is based on the parametric analysis by Chandy et al. (1975b). It is an iterative technique providing approximations for the queue length and waiting time distribution. Sauer and Chandy (1975) presented an approximate solution technique for flexible manufacturing systems needing less computation time. Their procedure is based on that of Chandy et al. They considered class-dependent service distributions and priority disciplines based on multiple-customer classes.

The approach of Shum and Buzen (1977) and Shum (1980) is called the extended product-form method. In their procedure, the Gordon-Newell Theorem is used for general processing times. The steady-state probabilities of the M/G/1 model were adapted according to the utilization that results from the estimated arrival rate.

Neuse and Chandy (1982) provided an approximate algorithm to analyze CQN with priority disciplines. It is referred to as the heuristic aggregation method (HAM)

---

[33]Curry and Feldman (2008, pp. 252ff).

Table 3.13 Approximate methods considering general processing times and infinite buffer space

Authors	C	M	S	D	BS	BM	R	BP	N	OA
Agrawal, Buzen, and Shum (1984)	CL	M	S	G	U	–	app	D	RTP	
Akyildiz (1987)	CR	M	S	ERL	U	–	app	MVA		[a]
Akyildiz and Sieber (1988)	CR	M	M	G	U	–	app	MM		[b]
Baynat and Dallery (1993)	CR	M	S	C	U	–	app	AT, MM		[c,d]
Baynat and Dallery (1996)	CR	M	S	G	U	–	app	MM		[c,d]
Baynat, Dallery, and Ross (1994)	CR	M	M	G	U	–	app	MM		[c]
Boxma and Donk (1982)	CL	2	S	1st G, 2nd E	U	–	app	LST		[e]
Cavaille and Dubois (1982)	CF	M	S	G	U	–	app	MVA		[c]
Chandy, Herzog, and Woo (1975a)	CR	M	S	G	U	–	app	NT		
Curry and Feldman (2008)	CR	M	S	G	U	–	app	MVA		
Dallery and Cao (1992)	CR	M	S	G	U	–	app	MM	OA	
Eager, Sorin, and Vernon (2000)	CL	M	S	HY	U	–	app	MVA		[f]
Gaver and Shedler (1973a)	CL	2	S	G	U	–	app	DA		
Gaver and Shedler (1973b)	CL	2	S	G	U	–	app	DA		
Gelenbe (1975)	CL	2	S	G	U	–	app	DA		
Kouvatsos and Almond (1988)	CL	2	S	G	U	–	app	ME		[g]
Marie (1979)	CR	M	S	G	U	–	app		MM	
Marie, Snyder, and Stewart (1982)	CR	M	S	G	U	–	app		MM	
Marie and Stewart (1977)	CR	M	S	G	U	–	app	MM		
Marie and Stewart (1983)	CR	M	S	G	U	–	app	MM		
Neuse and Chandy (1982)	CR	M	S	G	U	–	app	NT, MM	HAM, SCAT	[c]
Satyam and Krishnamurthy (2008)	CR	M	S	G	U	–	app	D		[c]
Sauer and Chandy (1975)	CF	M	S	G	U	–	app	NT		[c,h]
Schwerer and van Mieghem (1994)	CR	3	S	G	U	–	app	BM		[i]

(continued)

Table 3.13 continued

	C									
Shanthikumar and Gocmen (1983)	CR	M	S	LG	U	–	app	D		[j]
Shum (1980)	CR	M	S	G	U	–	app	NT, DA	EPF	
Shum and Buzen (1977)	CR	M	S	G	U	–	app	NT, DA	EPF	
Sun (2006)	CR	M	S	G	U	–	app	NT		
Whitt (1984)	CR	M	M	PH	U	–	app			[k]
Yao and Buzacott (1986a)	CF	S	S	G	U	–	app	MM		

Abbreviations of Table 3.13:

C configuration (*CF* flexible manufacturing system, *CL* closed and linear, *CR* closed and routing), *M* number of stations (*M* arbitrary many), *S* number of servers (*M* multiple, *S* single), *D* distribution of the processing times (*C* Coxian, *E* exponential, *ERL* Erlang, *G* general, *LG* load-dependent general, *HY* hyperexponential, *PH* phase-type), *BS* Buffer space (*L* limited, *U* unlimited), *BM* Blocking mechanism (*BAS* blocking after service, *BBS* blocking before service, *BR* block-and-recirculate), *R* result (*ex* exact, *app* approximate), *BP* base procedure (*AT* aggregation technique, *BM* Brownian Motion, *D* decomposition, *DA* Diffusion Approximation, *LST* Laplace Stieltjes Transformation, *ME* maximum entropy, *MM* Marie's method, *MVA* Mean value analysis, *NT* Norton's Theorem), *N* name of procedure (*EPF* Extended Product Form Solution, *MM* Marie's method, *OA* operational analysis, *RTP* Response time preservation, *SCAT* Self Correcting Approximation Technique), *OA* other aspects

[a] General processing times with: PS, LCFS-PR, IS, exponential processing times with: FCFS
[b] Subnetworks with population constraints
[c] Multiple-class customers
[d] Open systems with subnetworks having population constraints
[e] Closed-form expression
[f] High variability of the processing time distribution, $c^2 > 1$
[g] One station has multiple servers
[h] Service-by-priority discipline
[i] All stations same relative traffic intensity
[j] Lower and upper bounds for throughput
[k] Fixed population mean

or self-correcting aggregation technique (SCAT).[34] The method is based on Marie (1979) and Chandy et al. (1975b).

Boxma and Donk (1982) computed the cycle time distribution of two-station systems with one exponential and one general server via the Laplace transform. Kouvatsos and Almond (1988) analyzed two-station systems with multiple servers by the maximum entropy method. They proposed one-step recursions for the queue length distribution and established asymptotic relations to infinite capacity queues.

Agrawal et al. (1984) introduced the so-called response time preservation (RTP). In this procedure, a subsystem is replaced by an equivalent server whose response time in isolation equals that of the entire subsystem in isolation. The method is built upon the parametric analysis. However, it matches the cycle time rather than the throughput. In this method, general service times, multiple-customer classes, and priority scheduling are assumed.

Whitt (1984) investigated the relationship between open and closed queueing systems assuming phase-type distributed processing times. He proposed the idea of analyzing CQN by OQN with fixed population mean. Shanthikumar and Gocmen (1983) provided lower and upper bounds for the throughput of CQN with routing and load-dependent servers of arbitrary distribution. Using operational analysis, Dallery and Cao (1992) provided alternative derivations of some classical results regarding stochastic queueing networks. Further, they investigated the validity of product-form solutions, the aggregation property, and Marie's method.

Baynat and Dallery (1993) considered closed queueing networks with Coxian distributed service times and subnetworks having population constraints. They applied both the aggregation technique and Marie's method and compared the two approaches. Baynat et al. (1994) extended the method of Baynat and Dallery to multi-class networks with multiple servers. Baynat and Dallery (1996) considered CQN with multi-class customers and general processing times. They converted these into a single-class CQN with load-dependent service times. Each station was analyzed in isolation by a state-dependent Markovian process.

Cavaille and Dubois (1982) proposed an approximate mean value analysis (AMVA) for FMS with nearly deterministic processing times. They developed an estimate for the mean cycle time under the assumption of general processing times and multiple customer classes. According to Yao and Buzacott (1986a), their method is not adequate for service times with $c^2 > 1$. Yao and Buzacott developed the so-called exponentialization approach for multiple servers. Their approach is to transform the network with general processing times into an equivalent network with exponential state-dependent service times. They considered dynamic parts routing[35] and limited buffers.

Akyildiz (1987) proposed the so-called α-MVA for Erlang-k distributed service times. He derived an estimate for the mean residence time. Akyildiz and Sieber

---

[34]See also Sect. 3.3.1 on pages 26ff.

[35]Dynamic parts routing means that parts follow a probabilistic shortest-queue scheme. For details, see Yao and Buzacott (1985).

(1988) developed an approximate method for CQN with load-dependent general service time distributions. A new formula for the conditional throughput was used in an extension of Marie's method. Eager et al. (2000) modeled the service time distributions by means of a hyper-exponential distribution. They proposed an approximation of the mean residual service time and a new AMVA technique for computing the mean cycle time.

Bolch et al. (2006) applied the M/G/1 approximation for the mean work-in-process proposed by Kleinrock (1975) and extended several approximate approaches for exponential service times and infinite buffer capacities. One of their methods is called the Extended Bottapprox Method (EBOTT)[36] and is based on the Bottapprox approach by Bolch and Fischer (1993).[37] Further, they applied the M/G/1 formula to the procedure SUM, which is called the Extended Summation Method (ESUM). Bolch et al. moreover, introduced the so-called Extended Self-Correcting Approximation Technique (ESCAT) which is based on the approximate procedure for exponential CQN with infinite capacities called SCAT.[38]

Schwerer and van Mieghem (1994) studied three-station CQN with balanced workloads. They approximated the queue length by a reflected Brownian motion, resulting in closed-form solutions for the performance measures. Sun (2006) proposed a method for CQN with routing which is based on the procedure by Chandy et al. (1975b). However, in that approach, each node is analyzed in isolation. For the isolated analysis, a $\lambda(n)/G/1/N$ model was developed. The method particularly focuses on low coefficients of variation. Satyam and Krishnamurthy (2008) proposed a decomposition approach for multi-class CONWIP-systems. They used fork/join stations to model the synchronization of raw materials and CONWIP cards.

3.3.4 General Distribution and Finite Buffer Capacity

There are only a few methods fulfilling the assumption of generally distributed processing times and finite buffer capacities in closed queueing networks. In Table 3.14, an overview of these methods is given.

Akyildiz (1988a) proposed the so-called throughput analysis. A non-blocking network is fitted which approximates the performance measures of the blocking network. If the processing times follow a non-exponential distribution, Marie's method is applied for the evaluation of the corresponding non-blocking network.

---

[36]See Bolch et al. (2006, pp. 505ff).

[37]Bottapprox is an approximate method for CQN with exponential processing times and infinite buffers, see Sect. 3.3.1 on page 26.

[38]See Bolch et al. (2006, pp. 502–505).

Table 3.14 Approximate methods considering general processing times and finite buffer capacity

Authors	C	M	S	D	BS	BM	R	BP
Akyildiz (1988a)	CR	M	S	PH	L	BAS	app	α-MVA, MM
Bouhchouch, Frein, and Dallery (1996)	CL	M	S	G	L	BAS	app	D
Dayar and Meri (2008)	CR	M	S	PH	L	BAS	app	D, MC
Helber, Schimmelpfeng, and Stolletz (2011)	CL	M	S	G	L	BAS	app	LP
Matta and Chefson (2005)	CL	M	S	G	L	BBS	app	LP
Rall (1998)	CR	M	S	G	L	BAS	app	MM, MSS

Abbreviations:

C Configuration (CL closed and linear, CR closed and routing), M Number of stations (M arbitrary many), S number of servers (S single), D Distribution of the processing times (G general, PH phase-type), BS Buffer space (L limited), BM Blocking mechanism (BAS blocking after service, BBS blocking before service), R Result (app approximate), BP base procedure (D decomposition, LP linear programming, MC Markov chains, MM Marie's method, MSS Matching State Space)

In case of the exponential distribution, the α-MVA is used for the evaluation.[39] The method of Akyildiz is exact for two-station systems.

Rall (1998) used Marie's method for general CQN with infinite buffers and the matching-state-space method by Akyildiz (1987) for exponential CQN with finite buffers.[40] He transformed the network into a non-blocking network using Akyildiz's method and neglected the coefficient of variation. In the next step, he introduced a correction function subject to the coefficient of variation and the buffer capacity. This function is found by a linear regression applied to selected test instances. The regressors (the exogenous variables) constituted the analytical results given by Marie's method (of the general infinite-buffer CQN) and the regressands (the endogenous variables) were the simulation results (of the general finite-buffer CQN).[41]

Matta and Chefson (2005) developed a linear program that models CQN as a discrete-event system in continuous time. This way, the approximation quality corresponds to simulation. They further derived structural properties from the analysis of production lines with limited buffer capacities and stochastic processing times. Based on their approach, Helber et al. (2011) proposed a linear program in discrete time. The discrete time horizon and discrete state space allow the formulation of a buffer optimization model. Thereby, the performance analysis and the optimization is carried out simultaneously.

Dayar and Meri (2008) proposed a decomposition approach for CQN with phase-type distributed service times. The subnetworks are modeled by Markov chains which are represented by Kronecker products. The linkage of the subsystems is carried out by a combination of Marie's method and Yao and Buzacott's[42] method.

---

[39]See Akyildiz (1987) for details and page 41.

[40]See Akyildiz (1987, 1988c, 1989).

[41]See Rall (1998, p. 127).

[42]See Yao and Buzacott (1986a).

Based on the model by Bouhchouch et al. (1993, 1996) proposed a decomposition approach for general finite-buffer CQN. They analyzed an equivalent open queueing network with a virtual arrival rate, which has to be found. The subsystems were modeled by a continuous-time Markov chain and solved as described in Altiok and Ranjan (1989). An extension of this approach is presented in Chap. 4.

Chapter 4
Decomposition Approach

In the following, we present an iterative algorithm that is based on decomposition. It approximates the throughput of a closed queueing system with linear flow, general processing times, and finite buffer spaces. The service times at each station are assumed to be i. i. d. random variables with a general distribution that is described by the mean and the coefficient of variation. The stations consist of single servers with a first-come first-served queueing discipline. The results are obtained within seconds, and the approximation deviates on average by 2.43 % from simulation results over all test instances. The notation used in this chapter is provided in Table 4.1.

This chapter starts with the description of the algorithm in Sect. 4.1. This section is divided into the two parts that the algorithm consists of, namely the outer loop presented in Sect. 4.1.1 and the inner loop that is explained in Sect. 4.1.2. Thereupon, the method is examined by means of a numerical study in Sect. 4.2.

4.1 Algorithm

The inner and the outer loop each build upon an existing method. For the inner loop, we use the decomposition approach ALRMGEN by Manitz (2005) which analyzes open queueing systems with a linear flow of material, general processing times and finite buffers. In this approach, each subsystem (i.e., two work stations and their intervening buffer) is analyzed as a G/G/1/K stopped-arrival queueing system as proposed by Buzacott, Liu, and Shanthikumar (1995). The outer loop goes back to the procedure proposed by Bouhchouch, Frein, and Dallery (1993).

The virtual arrival rate to the open network is sought so that the estimated average work-in-process in the open network equals the given number of workpieces n in the closed queueing network under consideration. The resulting production rate of the open queueing network serves as the production-rate estimate of the closed queueing network.

S. Lagershausen, *Performance Analysis of Closed Queueing Networks*, Lecture Notes in Economics and Mathematical Systems 663, DOI 10.1007/978-3-642-32214-3_4, © Springer-Verlag Berlin Heidelberg 2013

Table 4.1 Notation

$C_{i,j}$	Buffer capacity between stations i and j
c_i	Coefficient of variation of the service time at station i
$\zeta_d(i,j)$	Coefficient of variation of the holding time at the downstream station in subsystem (i,j)
$\zeta_u(i,j)$	Coefficient of variation of the virtual inter-arrival time at the upstream station in subsystem (i,j)
λ	Virtual arrival rate to an open queueing network that is equivalent to the CQN being considered
$\lambda_{\min}$	Lower bound for λ
$\lambda_{\max}$	Upper bound for λ
M	Number of work stations in the network
μ_i	Service rate at station i
$\mu_u(i,j)$	Virtual arrival rate to the subsystem (i,j)
$\mu_d(i,j)$	Virtual service rate in the subsystem (i,j)
n	Number of carriers/pallets circulating in the closed queueing network
$p_S(i,j)$	Time-average starving probability in subsystem (i,j)
$p_B(i,j)$	Time-average blocking probability in subsystem (i,j)
$p_S^*(i,j)$	Arrival instant starving probability in subsystem (i,j)
$p_B^*(i,j)$	Arrival instant blocking probability in subsystem (i,j)
$L(i,j)$	Expected work-in-process in the subsystem (i,j) of the equivalent open queueing network
$\mathcal{P}_j$	Set of the predecessor stations of work station j
$PR(n)$	Production rate with n carriers circulating in the closed queueing network
T_i	Random variable describing the service time distribution at station i
$\mathcal{S}_i$	Set of the successor stations of work station i; $\mathcal{S}_M$ is the set of input stations that receive production-authorization information from the final work station M. With our assumptions, it holds that $\mathcal{S}_M = \{1\}$
$WIP(\lambda)$	Expected work-in-process in the open queueing network for a given virtual arrival rate λ
$X(\lambda)$	Production rate of the equivalent open queueing network for a given virtual arrival rate λ

Figure 4.1 provides an overview of the complete procedure. In the outer loop, a value for the virtual arrival rate λ is assigned.[1] Next, the work-in-process, $WIP(\lambda)$, of the equivalent open queueing system and the production rate, $X(\lambda)$, are evaluated in the inner loop.[2] Subsequently, in the outer loop, the $WIP(\lambda)$ is compared to the given number of workpieces n (which represents the desired work-in-process). If $WIP(\lambda)$ and n differ, the value of λ is systematically changed and the equivalent open system is reevaluated. This is repeated until the values of $WIP(\lambda)$ and n are close. If so, the algorithm terminates and the current production-rate estimate of the open system serves as the estimate for that of the closed queueing system.

---

[1] The assignment of λ is carried out according to a mechanism that is explained in Sect. 4.1.1.
[2] The evaluation is explained in Sect. 4.1.2.

Fig. 4.1 Flowchart of the
overall approach

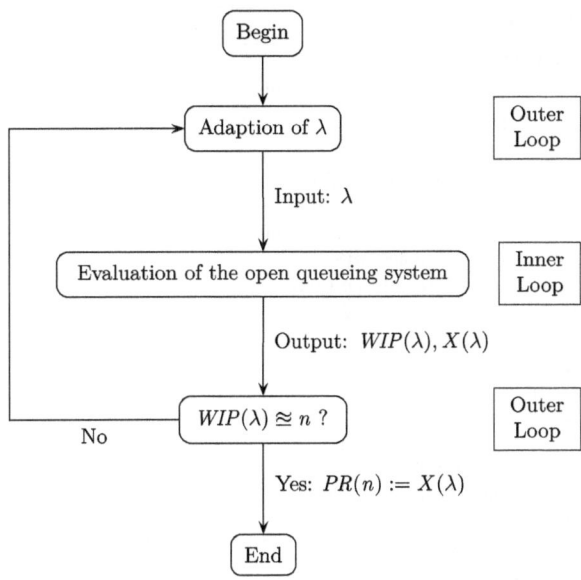

In summary, in the outer loop, the arrival rate λ is adapted. In the inner loop, the corresponding open queueing network with arrival rate λ is evaluated.

4.1.1 Outer Loop

The purpose of the outer loop is to find the artificial arrival rate λ to an open system that leads to a work-in-process of n. The search mechanism consists in narrowing down an upper and a lower bound. The upper bound for λ is initialized with the lowest processing rate among all processing stations, i.e., $\lambda_{max} := \min_{m \in \{1,\dots,M\}} \{\mu_m\}$. The value of that upper bound is explained as follows: A change of the arrival rate between two values that are higher than the smallest processing rate does not change the production rate. Therefore, the lowest processing rate represents the upper bound for an arrival rate which influences the production rate. That initialization is tighter than the initialization of Bouhchouch et al. (1993), and therefore, accelerates the calculation as compared to their approach (which would be $\lambda_{max}^{\mathrm{BFD}} := \mu_M$). The lower limit of λ is simply set at 0: $\lambda_{min} := 0$.

In each iteration, the arrival-rate estimate λ is determined as the average of its current upper and lower bound:

$$\lambda := \frac{\lambda_{\min} + \lambda_{\max}}{2}. \tag{4.1}$$

Algorithm 4 Pseudo-code of the outer loop

1: **procedure** OUTER LOOP
2:      $\lambda_{\min} := 0, \lambda_{\max} := \min_{m \in \{1,...,M\}} \{\mu_m\}$
3:      **repeat**
4:          $\lambda := \frac{\lambda_{min} + \lambda_{max}}{2}$
5:          **procedure** ALRMGEN(λ)
6:              **do** INNER LOOP
7:              **return** WIP(λ), $X(\lambda)$
8:          **end procedure**
9:          **if** $WIP(\lambda) > n$ **then**
10:             $\lambda_{max} := \lambda$
11:         **else**
12:             $\lambda_{min} := \lambda$
13:         **end if**
14:         **if** $\lambda_{max} - \lambda_{min} < \epsilon$ **then**
15:             Exit repeat
16:         **end if**
17:     **until** $|WIP(\lambda) - n| \le \epsilon$
18:     $PR(n) := X(\lambda)$
19:     **return** PR(n)
20: **end procedure**

For a given value of λ, the inner-loop algorithm is started. In the inner loop, the work-in-process approximation WIP(λ) and the production-rate estimate $X(\lambda)$ are calculated.[3] The work-in-process estimate WIP(λ) in comparison to n implies either a new lower or a new upper bound for λ. The adaption of the bounds is based upon the following coherence between arrival rate and work-in-process: If the arrival rate increases, the work-in-process also increases (or stays the same) and vice versa. Therefore, if WIP(λ) exceeds n, the virtual arrival rate λ has been pre-specified too high. In this case, it can be stated that the current value of the arrival rate serves as new upper bound. Consequently, the upper bound λ_{max} is set equal to the current value of λ: $\lambda_{\max} := \lambda$. By applying (4.1), λ is lowered. If WIP(λ) is smaller than n, the arrival rate is too low and the current value for λ serves as the new lower bound: $\lambda_{\min} := \lambda$. In that case, λ is increased by utilization of Eq. (4.1).

With the updated virtual arrival rate λ, the performance measures of the equivalent open network are reevaluated in the inner loop. Thereafter, the work-in-process estimate is again compared to the number of workpieces n, and, if necessary, λ is adapted again. This is carried out until the difference between n and WIP(λ) lies within a pre-specified tolerance of $\epsilon = 0.000001$. The procedure of the outer loop is summarized in Algorithm 4.

The development of the values of λ, $\lambda_{\min}$, and $\lambda_{\max}$ can be reconstructed from Fig. 4.2. For a five-station system with a linear flow of material, the course of the values can be traced throughout one run of the algorithm, once for $n = 1$ and once for $n = 10$. The data of the CQN being considered are provided in Table 4.2.

---

[3]This is explained in Sect. 4.1.2.

Fig. 4.2 The values of λ, $\lambda_{\min}$, and $\lambda_{\max}$ for subsequent outer-loop iterations; $n = 1$ (*top*), and $n = 10$ (*bottom*)

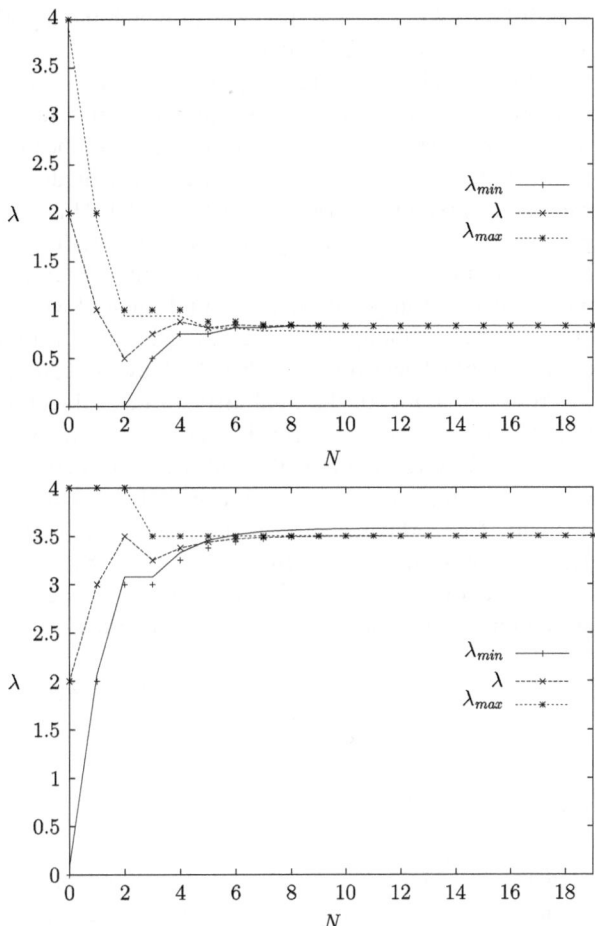

Table 4.2 Example instance of a five-station system

m	1	2	3	4	5
μ_m	4	5	4	5	4
c_m	0.5	0.6	0.5	0.6	0.7
Buffer size	$C_{5,1} = 4$	$C_{1,2} = 4$	$C_{2,3} = 4$	$C_{3,4} = 4$	$C_{4,5} = 4$

As one can see, the gap between the upper and lower bound is narrowed down iteratively. The values for λ are found exactly in the middle of the current bounds.

In a few cases, WIP(λ) does not approach n with an increasing number of iterations.[4] To avoid an infinite search for an appropriate λ and to ensure termination, the algorithm stops when $\lambda_{\min}$ and $\lambda_{\max}$ are very close (i.e. if $\lambda_{\max} - \lambda_{\min} \leq 0.000001$),

---

[4]For more information, see Sect. 4.1.2.

even if $|\text{WIP}(\lambda) - n| > \epsilon$. In cases in which the algorithm does not terminate, the production-rate estimate is, therefore, poor.

Figure 4.3 shows the final values for λ for different numbers n of circulating carriers in the closed-loop line running from 1 to 24. As one can see, the final values for λ become larger with a larger number of carriers in the system, i.e. with a higher WIP level. The saddle in the medium range indicates the interval where the maximum production rate is found. This can be verified using the bottom graph in Fig. 4.3 by analyzing the plot of the production-rate estimates as a function of n. With increasing n, the system becomes saturated and λ approaches its upper bound. In this example, the upper bound equals 4 which conforms with the minimum processing rate among the processing stations. From a certain n on, an increasing number of blockages due to higher WIP levels decreases the production rate.[5]

From the course of the final arrival rates subject to n, $\lambda(n)$, it follows that the initial lower bound for λ can be set to a tighter value than zero when iterating over all n. The reasoning is the following: To achieve a work-in-process of $n + 1$ workpieces in the open system, the arrival rate $\lambda(n + 1)$ must be at least as high as it is with one workpiece fewer in the system, $\lambda(n)$ (all other parameters kept equal). That means, it holds that $\lambda(n + 1) \geq \lambda(n) \; \forall n < N$. Therefore, when iterating over n, the lower bound for the arrival rate, λ_{min}, can be set at the value of the final arrival rate λ from the former iteration, see Eq. (4.2).

$$\lambda_{min}(n + 1) := \lambda(n) \qquad\qquad (4.2)$$

4.1.2 Inner Loop

In the inner loop, a pseudo-open queueing network with an arrival rate of λ is analyzed. The analysis is based on the decomposition approach ALRMGEN proposed by Manitz (2008). The procedure ALRMGEN iteratively considers two-station subsystems with a finite buffer inbetween.

In Sect. 4.1.2.1, the adapted decomposition approach is presented, which is followed by the estimation of the work-in-process in Sect. 4.1.2.2.

4.1.2.1 Adapted ALRMGEN

In contrast to the usual assumptions for saturated open queueing networks, the final station may be blocked, and the input station may be starved. This is taken into account by an additional subsystem which is considered within the decomposition approach. In that subsystem, the final processing station indicates the upstream

---

[5]For further details on the coherence of the production-rate function and the blocking and starving probabilities, see Sect. 2.2.

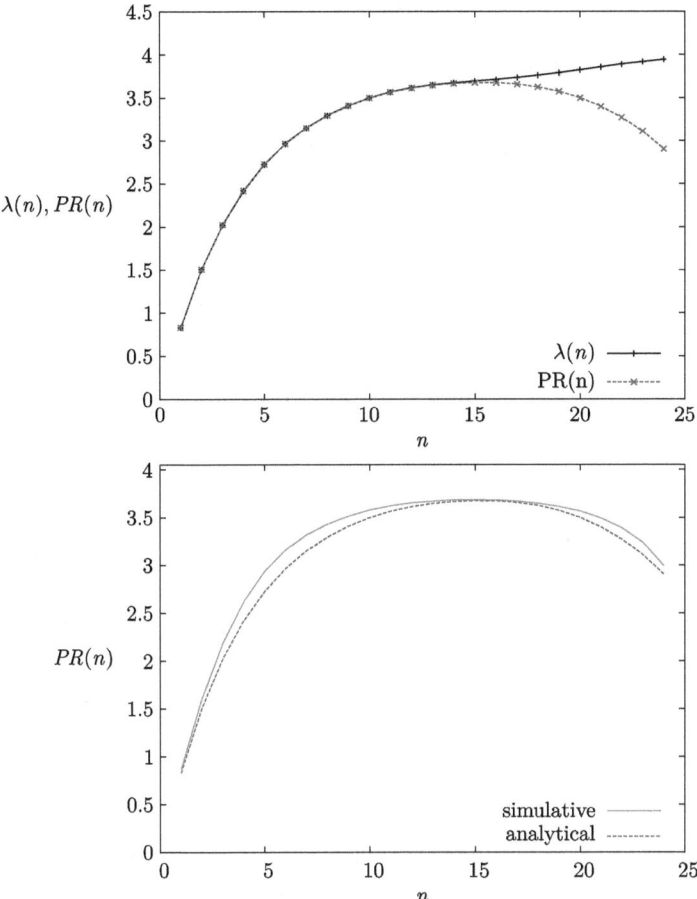

Fig. 4.3 The final estimates for the virtual arrival rate λ (*top*) and for the production rate (*bottom*)

station, and the input station mimics the downstream station. Figure 4.4 displays the subsystems of the upstream and downstream pass of a four-station system.

In place of the first station, a station is installed whose parameters differ depending on whether the upstream or the downstream pass is performed. For the upstream pass, the first station mimics the arrival process, given by the arrival rate λ and the coefficient of variation $\zeta_u(1, 2)$. The parameters of that station are never updated, thus enabling the first station to control the work-in-process of the line. In the downstream pass of subsystem $(1, 2)$, the processing time distribution corresponds to that of station 1 in the closed system. The blocking of the last station and the starving of the first station are captured in the additional subsystem $(M, 1)$.

Throughout the inner loop, we set the rate of the input station equal to the virtual arrival rate that has been pre-specified in the outer loop, see Eq. (4.3).

$$\mu_u(1, 2) = \lambda \tag{4.3}$$

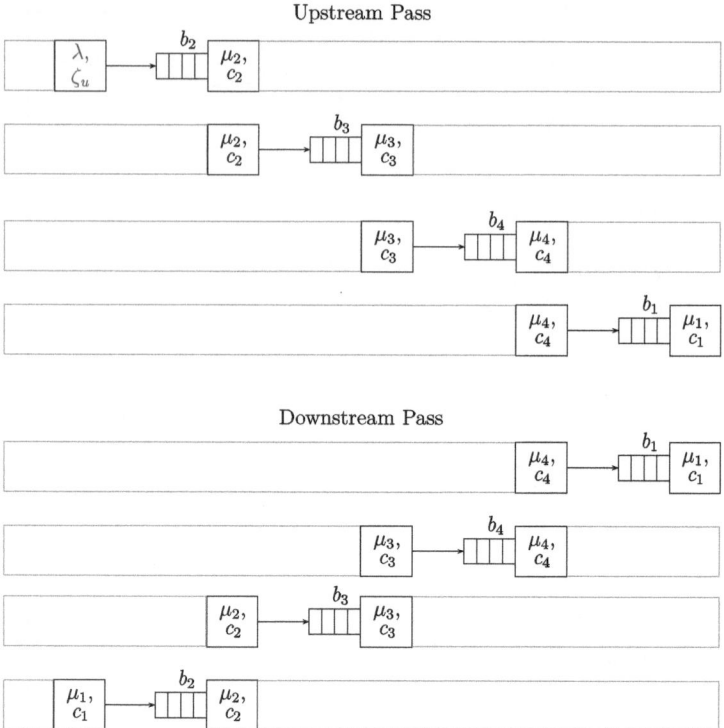

Fig. 4.4 Subsystems

The virtual arrival process is supposed to have no variability. Variability in the arrival process would only increase the variability of the work-in-process in the system. A more variable work-in-process would lead to less accurate results in the production-rate estimate.[6] Therefore, the coefficient of variation for the inter-arrival time at the input station is set to:

$$\zeta_u(1, 2) := 0 \qquad (4.4)$$

The other parameters of the inner loop, namely the time-average starving and blocking probabilities $p_S(i, j)$ and $p_B(i, j)$, the arrival instant starving and blocking probabilities $p_S^*(i, j)$ and $p_B^*(i, j)$, the virtual arrival and service rates $\mu_u(i, j)$ and $\mu_d(i, j)$, the coefficient of variation of the holding time at the downstream station $\zeta_d(i, j)$, and the virtual inter-arrival time at the upstream station $\zeta_u(i, j)$, each for subsystem (i, j), are initialized as follows:

---

[6]Numerical tests with different $\zeta_u(1, 2)$ proved that the mean deviation of the production-rate estimate over all test instances is higher, the higher the coefficient of variation of the inter-arrival time at the input station, $\zeta_u(1, 2)$, is.

$$p_S(i, j) = 0, \quad p_S^*(i, j) = 0 \qquad (i \in \{1, \ldots, M\}; j \in \mathcal{S}_i) \qquad (4.5)$$

$$p_B(i, j) = 0, \quad p_B^*(i, j) = 0 \qquad (i \in \{1, \ldots, M\}; j \in \mathcal{S}_i) \qquad (4.6)$$

$$\mu_u(i, j) = \mu_i, \quad \zeta_u(i, j) = c_i \qquad (i \in \{2, \ldots, M\}; j \in \mathcal{S}_i) \qquad (4.7)$$

$$\mu_d(i, j) = \mu_j, \quad \zeta_d(i, j) = c_j \qquad (i \in \{1, \ldots, M\}; j \in \mathcal{S}_i) \qquad (4.8)$$

The upstream station of a subsystem generates the arrival process of parts into the buffer and represents the overall effects of the arrival process. We denote a particular subsystem by (i, j). The indices i and j are those of the two stations that are involved, with j being the successor station of i, denoted by $j \in \mathcal{S}_i$. The arrival process of parts that is generated by processing parts at the upstream station is described by the rate $\mu_u(i, j)$ and the coefficient of variation of the inter-arrival time, $\zeta_u(i, j)$. The downstream station j empties the buffer by processing parts. This process is described by the (virtual) processing rate $\mu_d(i, j)$ and the coefficient of variation of the processing time, $\zeta_d(i, j)$ (with i as a predecessor station of $j : i \in \mathcal{P}_j$). It reflects the effective holding time of a part at station j.

The procedure of the inner loop is summarized in Algorithm 5. With the given virtual arrival rate λ, an inner loop iteration begins. For this λ, the performance of the equivalent open network is evaluated with the initializations according to Eqs. (4.5)–(4.8). The upstream parameters are updated for each station except for the first station. In the downstream pass, the parameters of all subsystems are updated. With the updated parameters, the current values of the probabilities of starvation and blockage can be calculated and, with them, the rates and coefficients of the variation of the upstream and downstream station are updated. Thereupon, these parameters are used in the next iteration.

The ALRMGEN procedure terminates when the estimate of the production rate converges to a stable value. This value is denoted by $X(\lambda)$. The production-rate estimate of the open system, $X(\lambda)$, in the last iteration of the outer loop, represents the production-rate estimate of the equivalent closed queueing system, $PR(n)$.

4.1.2.2 Work-in-Process Estimation

When the ALRMGEN procedure has terminated, the work-in-process estimate $WIP(\lambda)$ is calculated by summing up the expected work-in-process of subsystem (i, j), $L(i, j)$, over all subsystems, see Eq. (4.9).

$$\text{WIP}(\lambda) = \sum_{i=1}^{M} L(i, j) \qquad j \in \mathcal{S}_i. \qquad (4.9)$$

The expected number of workpieces $L(i, j)$ in the subsystem (i, j) is derived by weighing the number of n_j workpieces with the steady-state probabilities for n_j

---

Algorithm 5 Inner loop

1: **procedure** INNER LOOP(λ, μ_i, c_i, $C_{i,j}$ $\forall i$)
2:      $j := S_i$ $\forall i$
3:      **for** $i = 1$ **to** M **do**   ▷ Initialization
4:          **if** $i = 1$ **then**
5:              $\mu_u(i, j) = \lambda$
6:              $\zeta_u(i, j) = 0.01$
7:          **else if** $i > 1$ **then**
8:              $\mu_u(i, j) = \mu_i$
9:              $\zeta_u(i, j) = c_i$
10:         **end if**
11:         $\mu_d(i, j) = \mu_j$
12:         $\zeta_d(i, j) = c_j$
13:     **end for**
14:     **repeat**
15:         **for** $i = 1$ **to** M **do**   ▷ Upstream Pass
16:             **if** $i \neq 1$ **then**
17:                  Update μ_u, ζ_u
18:             **end if**
19:             Update $p_S, p_B, X_u, X_d, p_S^*, p_B^*$
20:         **end for**
21:         **for** $i = M$ **down to** 1 **do**   ▷ Downstream Pass
22:             Update μ_u, ζ_u
23:             Update $p_S, p_B, X_u, X_d, p_S^*, p_B^*$
24:         **end for**
25:         $X(n) = \mu_d(M - 1, M) \cdot (1 - P_S(M - 1, M))$
26:     **until** $|X^{(new)} - X^{(old)}| \leq \delta$
27:     **procedure** WIP-LEVEL_SUBSYSTEM($\mu_u, \zeta_u, \mu_d, \zeta_d, C$)
28:         **return** $L(i, j)$ $\forall i, j \in S_i$
29:     **end procedure**
30:     $WIP(\lambda) = \sum_{i=1}^{M} L(i, j)$
31:     **return** WIP(λ), $X(\lambda)$
32: **end procedure**

---

workpieces at station j, $P(n_j)$, for all $n_j = 1, \ldots, C_{i,j} + 1$, $j \in S_i$, as shown in Eq. (4.10).

$$L(i, j) = \sum_{n_j=1}^{C_{i,j}+1} n_j \cdot P(n_j), \qquad \forall i, j \in S_i \qquad (4.10)$$

The approximation formulas of the steady-state probabilities for n_j workpieces at station j, $P(n_j)$, are obtained from the G/G/1/K stopped-arrival queueing model proposed by Buzacott and Shanthikumar (1993). The input for the calculation of $P(n_j)$ are the final upstream and downstream parameters μ_u, ζ_u, μ_d, ζ_d, and Z. Z denotes the station capacity plus an additional position for the case of blocking,

---

Algorithm 6 Approximation of the expected number of jobs in an open $G/G/1/K$ queueing system with stopped arrivals

---

1: **function** WIP-LEVEL_SUBSYSTEM$(i, j)(\mu_u, \zeta_u, \mu_d), \zeta_d, C)$

2: $\quad \rho = \dfrac{\mu_u(i, j)}{\mu_d(i, j)}$

3: $\quad$ **if** $\rho \neq 1$ **then**

4: $\quad\quad$ **if** $\rho < 1$ **then**

5: $\quad\quad\quad \widehat{L} = \dfrac{\rho^2 \cdot (\zeta_d^2(i, j) + 1)}{1 + \rho^2 \cdot \zeta_d^2(i, j)} \cdot \dfrac{\zeta_u^2(i, j) + \rho^2 \cdot \zeta_d^2(i, j)}{2 \cdot (1 - \rho)} + \rho$

6: $\quad\quad$ **else** $(\rho > 1)$

7: $\quad\quad\quad \widehat{L} = \dfrac{\dfrac{1}{\rho^2} \cdot (\zeta_u^2 + 1)}{1 + \dfrac{1}{\rho^2} \cdot \zeta_u^2} \cdot \dfrac{\zeta_d^2 + \dfrac{1}{\rho^2} \cdot \zeta_u^2}{2 \cdot \left(1 - \dfrac{1}{\rho}\right)} + \dfrac{1}{\rho}$

8: $\quad\quad$ **end if**

9: $\quad\quad \sigma = \dfrac{\widehat{L} - \rho}{\widehat{L}}$

10: $\quad\quad L(i, j) = \dfrac{\rho \cdot \sigma \cdot (\sigma^{C_{i,j} + 2} - 1) + (C_{i,j} + 2) \cdot \rho^2 \cdot \sigma^{C_{i,j} + 2} \cdot (1 - \sigma)}{\sigma \cdot (1 - \sigma) \cdot (\rho^2 \cdot \sigma^{C_{i,j} + 1} - 1)}$

11: $\quad$ **else**

12: $\quad\quad L(i, j) = \dfrac{C_{i,j} + 2}{2}$

13: $\quad$ **end if**

14: **end function**

$Z = C_{i,j} + 2$.[7] The calculation of the steady-state probabilities depends on the traffic intensity $\rho = \frac{\mu_u}{\mu_d}$.[8] The formulas of the work-in-process estimate for the subsystem (i, j), $L(i, j)$, are derived by inserting the expressions for the $P(n_j)$ into Eq. (4.10) and rearranging the expressions. The result is displayed in Algorithm 6.

As mentioned above, in a few cases, the decomposition approach does not yield satisfying results. The bisection search described in Sect. 4.1.1 assumes implicitly that the work-in-process estimate is a monotone increasing function subject to λ (and thereby subject to ρ). From the theory, it is straightforward that an increase in the arrival rate (and thereby in the utilization) leads to an increase (or same level) of the work-in-process. However, in the decomposition approach, this is not always the case for large n. To accomplish a high work-in-process, the arrival rate must also be high relative to the processing rates. This means that the utilization in the

---

[7] The station capacity equals the buffer capacity, $C_{i,j}$, plus one unit for the server.

[8] For the calculation of $P_i(n_i)$, see Manitz (2005, p. 138).

Table 4.3 Test instances taken from Bouhchouch et al. (1996)

Cases	n	M	$C_{1,2} \cdots C_{M,1}$	$\mu_1 \cdots \mu_M$	$c_1 \cdots c_M$
BFD96_Ex1	1–19	5	3 3 3 3 3	1 1 1 1 1	1 1 1 1 1
BFD96_Ex2a	1–41	7	5 5 5 5 5 5 5	1 1 1 1 1 1 1	$\sqrt{5}$ $\sqrt{5}$ $\sqrt{5}$ $\sqrt{5}$ $\sqrt{5}$ $\sqrt{5}$ $\sqrt{5}$
BFD96_Ex2b	1–41	7	5 5 5 5 5 5 5	1 1 1 1 1 1 1	$\sqrt{2}$ $\sqrt{2}$ $\sqrt{2}$ $\sqrt{2}$ $\sqrt{2}$ $\sqrt{2}$ $\sqrt{2}$
BFD96_Ex2c	1–41	7	5 5 5 5 5 5 5	1 1 1 1 1 1 1	1 1 1 1 1 1 1
BFD96_Ex2d	1–41	7	5 5 5 5 5 5 5	1 1 1 1 1 1 1	$\sqrt{0.5}$ $\sqrt{0.5}$ $\sqrt{0.5}$ $\sqrt{0.5}$ $\sqrt{0.5}$ $\sqrt{0.5}$ $\sqrt{0.5}$
BFD96_Ex3	1–45	7	7 4 6 5 6 7 4	1 2 1.5 1 1.25 2 1.75	1 1 1 1 1 1 1
BFD96_Ex4	1–61	10	4 6 4 4 7 4 4 6 6 7 1	1.5 1 1 2 1 1 1.5 1.5 2	1 $\sqrt{0.5}$ $\sqrt{2}$ $\sqrt{5}$ $\sqrt{0.5}$ $\sqrt{1}$ $\sqrt{0.5}$ $\sqrt{2}$ $\sqrt{5}$ $\sqrt{5}$

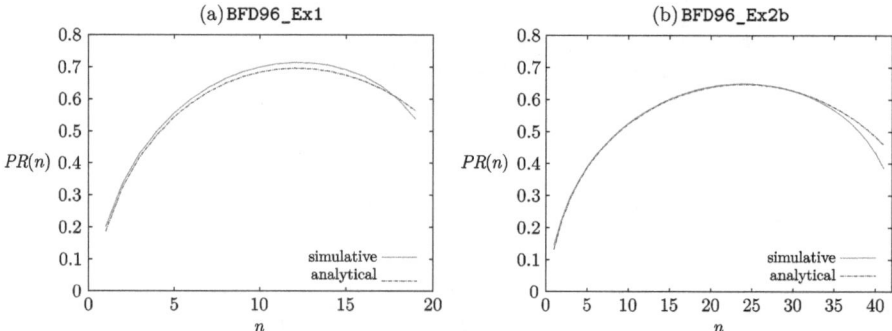

Fig. 4.5 Production rate results of selected instances. (**a**) BFD96_Ex1. (**b**) BFD96_Ex2b

subsystems is close to 1. As can be seen in Algorithm 6, there are different formulas for $L(i, j)$, depending on whether $\rho < 1$ or $\rho = 1$. Depending on the values, the work-in-process for $\rho = 1$ may be smaller than for a ρ that is (smaller but) close to 1. If so, the logic of adapting λ is leveraged and the algorithm fails. Then, $\lambda_{\min}$ and $\lambda_{\max}$ are very close and further iterations do not have an impact on the production-rate or the work-in-process estimate. To avoid an infinite loop that might occur in that situation, the outer-loop algorithm also terminates if $\lambda_{\max} - \lambda_{\min} \leq 0.00001$.

4.2 Numerical Study

In this section, we compare our approximation to examples reported in Bouhchouch, Frein, and Dallery (1996) and to simulation results using our own test sets. First, we investigate the performance of our algorithm compared to the one proposed by Bouhchouch et al. (1996). The input data are summarized in Table 4.3. The analytical results of our algorithm in comparison to simulation results are depicted in Fig. 4.5 for the Cases BFD96_Ex1 in graph (a) and BFD96_Ex2b in graph (b).

Table 4.4 Analytical results of Bouhchouch et al. (1996) and our approach

Case	n^*	$X(n^*)_{\text{Sim}}$	$X(n^*)_{\text{L}}$	$\frac{X(n^*)_{\text{L}}-X(n^*)_{\text{Sim}}}{X(n^*)_{\text{Sim}}}$ (%)	$X(n^*)_{\text{BFD}}$	$\frac{X(n^*)_{\text{BFD}}-X(n^*)_{\text{Sim}}}{X(n^*)_{\text{Sim}}}$ (%)
BFD96_Ex2a	25	0.4936	0.4791	−2.9	0.4570	−7.4
BFD96_Ex2b	24	0.6482	0.6470	−0.1	0.6282	−3.1
BFD96_Ex2c	25	0.7664	0.7640	−0.2	0.7486	−2.3
BFD96_Ex2d	25	0.8567	0.8561	−0.1	0.8464	−1.2

Table 4.5 Analytical results of Akyildiz (1988a) and our approach

c_1^2	c_2^2	$X(10)_{\text{exact}}$	$X(10)_{\text{L}}$	$\frac{X(10)_{\text{L}}-X(10)_{\text{exact}}}{X(10)_{\text{exact}}}$ (%)	$X(10)_{\text{A}}$	$\frac{X(10)_{\text{A}}-X(10)_{\text{exact}}}{X(10)_{\text{exact}}}$ (%)
0.5	0.5	1.9790	1.891632905	−4.41	1.9761	−0.15
0.5	1	1.9469	1.837981683	−5.59	1.9469	0.00
0.5	4	1.7769	1.571599853	−11.55	1.7992	1.25
0.5	9	1.6632	1.318247861	−20.74	1.6906	1.65
1	0.5	1.9402	1.835739281	−5.38	1.9402	0.00
1	1	1.9038	1.779883622	−6.51	1.9038	0.00
1	4	1.7510	1.527958331	−12.74	1.7510	0.00
1	9	1.6497	1.292207092	−21.67	1.6497	0.00

A comparison between the results of our algorithm and the results from Bouhchouch et al. (1996) is presented in Table 4.4. In their paper, numerical results are reported only for four particular configurations, each with the optimal CONWIP level n^*. The results for the production rate of our approach, denoted by $X(n^*)_{\text{L}}$, are compared both to the analytical results generated by Bouhchouch et al., denoted by $X(n^*)_{\text{BFD}}$, and to simulation results according to their paper, denoted by $X(n^*)_{\text{Sim}}$. As can be seen, our approach performs better in all the reported cases.

We further compare our approach to that of Akyildiz (1988a). He investigated the performance of two-station systems with $b_1 = 5$, $b_2 = 6$, $\mu_1 = 2$, $\mu_2 = 3$, $n = 10$ and different coefficients of variation. A comparison of these instances is presented in Table 4.5. It can be seen that our approach is not so valuable for very small queueing systems.

For a second series of numerical experiments, we set up our own test sets. We varied the processing rates μ, the coefficients of variation of the processing times c, and the buffer sizes C. For each parameter, cases were chosen to consider low (1), medium (m) (only for selected parameters), and high (h) values, and balanced (b) or unbalanced (u) configurations of the queueing system. We used the settings

$$\text{mu\_u\_l} = \{0.7, 0.8, 0.6, 0.5, 0.9, 0.8, 0.5, 0.7, 0.9, 0.6\}$$

$$\text{mu\_u\_m} = \{1.2, 1.5, 0.5, 0.7, 2, 0.9, 1.1, 1.2, 1.3, 0.9\}$$

$$\text{mu\_u\_h} = \{4, 5, 4, 5, 4, 5, 4, 5, 4, 5\}$$

$$\text{mu\_b\_h} = \{1, 1, 1, 1, 1, 1, 1, 1, 1, 1\}$$

for the processing rates $\mu_1, \ldots, \mu_{10}$. The CVs were

$$CV\_b\_l = \{0.25, 0.25, 0.25, 0.25, 0.25, 0.25, 0.25, 0.25, 0.25, 0.25\}$$
$$CV\_b\_h = \{1, 1, 1, 1, 1, 1, 1, 1, 1, 1\}$$
$$CV\_u\_l = \{0.5, 0.6, 0.5, 0.6, 0.5, 0.6, 0.5, 0.6, 0.5, 0.6\}$$
$$CV\_u\_h = \{1.2, 0.8, 1.5, 1.1, 0.9, 0.4, 0.6, 1.3, 0.8, 0.7\}$$

for $c_1, \ldots, c_{10}$. The buffer sizes $C_{10,1}, \ldots, C_{9,10}$ were set to

$$C\_b\_l = \{4, 4, 4, 4, 4, 4, 4, 4, 4, 4\}$$
$$C\_b\_h = \{12, 12, 12, 12, 12, 12, 12, 12, 12, 12\}$$
$$C\_u\_l = \{3, 6, 4, 5, 6, 3, 4, 5, 6, 5\}$$
$$C\_u\_h = \{8, 7, 12, 9, 11, 13, 8, 12, 10, 11\}.$$

The first five figures of each parameter represent the data for the five-station systems. For both the five-station and the ten-station line, every parameter value is combined with every other, which results in $4 \cdot 4 \cdot 4 \cdot 2 = 128$ test instances.

For the case mu_u_h-CV_u_h-C_b_l, numerical results for the production-rate estimates are depicted in Fig. 4.6, both for the five-station system in (a) and the ten-station system in (b). The results are plotted over the range of $n = 1, \ldots, N$. As one can see from both figures, the approximation procedure works very well. It works best for a medium range of the number n of jobs in the system. From Fig. 4.6b, it can be observed that the decreasing arc of the throughput curve is not fitted as well as in the middle of the range. However, the approximation error for $n > n^*$ is not relevant because all $n > n^*$ are suboptimal from an economic point of view. Over all test cases, the mean deviation of estimates from simulation results for $X(n^*)$ was 1.06 %. While the quality of the approximation is quite stable for the middle of the range, the few cases with large n, for which the approximation fails, were reflected in greater proportion to the mean deviation. However, this increase is rather moderate at 2.43 % indicating that the approximation gives good results for throughput estimates over the entire range of n.

For a final series, data for a 20-station systems were used. The test sets were set up as shown below. The processing rate settings $\mu_1, \ldots, \mu_{20}$ are

$$
\begin{aligned}
mu\_u\_l = \{ & 0.19, 0.19, 0.18, 0.17, 0.27, 0.18, 0.17, 0.19, 0.27, 0.18, \\
& 0.17, 0.18, 0.22, 0.28, 0.19, 0.18, 0.24, 0.19, 0.19, 0.17\}
\end{aligned}
$$

$$
\begin{aligned}
mu\_u\_m = \{ & 0.85, 0.85, 0.8, 0.85, 0.85, 0.9, 1, 1, 1, 1, \\
& 0.9, 0.85, 0.8, 0.8, 1, 1, 1, 1, 1, 1\}
\end{aligned}
$$

$$
\begin{aligned}
mu\_u\_h = \{ & 2.1, 2.1, 2, 2.05, 2.15, 2.15, 1.95, 2.1, 2.05, 2.05, \\
& 2, 2.1, 1.95, 2.205, 2.05, 2.3, 2.1, 2, 2.1, 2\}
\end{aligned}
$$

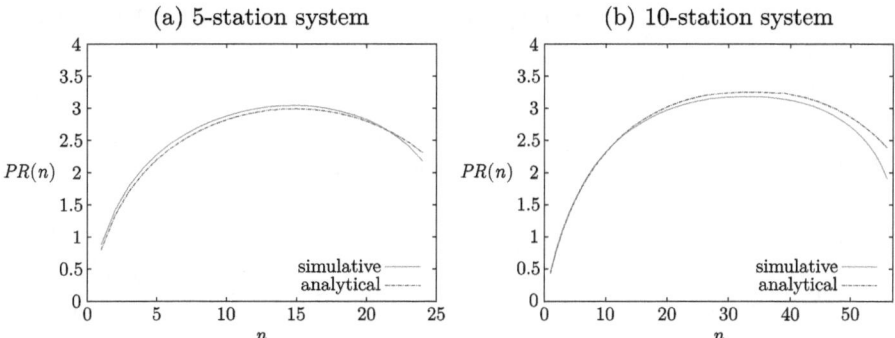

Fig. 4.6 Production rate results of case mu_u_h-CV_u_h-C_b_1. **(a)** Five-station system.
(b) Ten-station system

and the buffer sizes for $C_{20,1}, C_{1,2}, \ldots, C_{19,20}$ are

$$C\_u\_1 = \{1, 2, 1, 2, 1, 2, 1, 2, 1, 2, 1, 2, 1, 2, 1, 2, 1, 2, 1, 2\}$$
$$C\_u\_m = \{1, 3, 1, 5, 1, 4, 1, 3, 2, 1, 1, 3, 1, 5, 1, 4, 1, 3, 2, 1\}$$
$$C\_u\_h = \{5, 7, 6, 4, 6, 5, 7, 5, 4, 3, 5, 7, 6, 4, 6, 5, 7, 5, 4, 3\}.$$

The processing rates indicate an unbalanced system, while the differences are small.
The coefficients of variation of the processing times are assumed to be equal at
all stations. We use the values $c_i = 0.1$ (vl), 0.4 (l), 0.7 (m), and 1.2 (h) for
all i. The combination of all parameter sets leads to $3 \cdot 3 \cdot 4 = 36$ test cases.
As for the other cases, we first report two selected cases and then analyze the
mean deviations of the throughput estimates over all test cases for this data set.
The graphical representations of the selected cases are given in Fig. 4.7.

Again, the quality of the approximation procedure can be shown for a wide range
of n in the closed queueing system. In the middle range of n, the approximation
works very well; for large and small values of n, the approximation does not. One
can see that the approximation fails in the case where—during the algorithm's
run—no λ can be found that yields a WIP(λ) that is close to n. That is the
effect explained in Sect. 4.1.2. The critical value of Case mu_u_h-CV_b_m-C_u_h is
$n = 90$. However, up to this value, the approximation is fairly accurate as compared
to simulation results.

In Table 4.6, the mean relative deviations of test set 1 (5- and 10-station systems)
and test set 2 (20-station systems) are summarized, each for $n = 1, \ldots, N$ and the
production rate maximal n^*. The mean relative deviation is calculated by

$$MRD(PR) = \left| \frac{PR_{proc} - PR_{sim}}{PR_{sim}} \right|. \qquad (4.11)$$

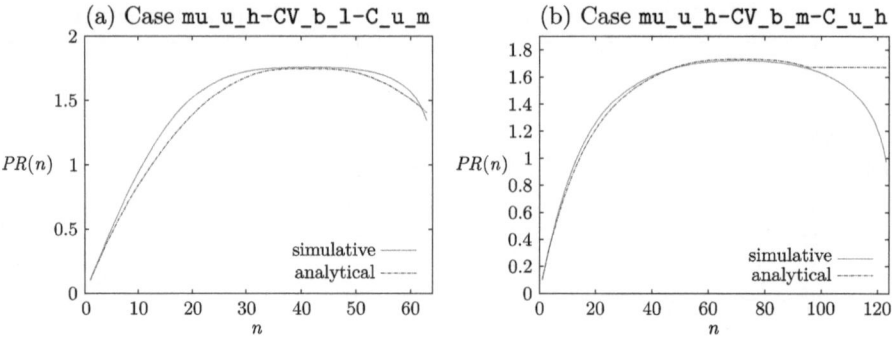

Fig. 4.7 Production rate results for the 20-station systems. **(a)** Case mu_u_h-CV_b_1-C_u_m. **(b)** Case mu_u_h-CV_b_m-C_u_h

Table 4.6 Mean relative deviations of test set 1 and test set 2

	$n = 1, \ldots, N$ MRD(PR) (%)	n^* MRD(PR) (%)
Test set 1	2.43	1.06
Test set 2	4.7	0.06

In test set 2, the mean relative deviation of the production-rate estimates from simulation results is 0.06 % for n^*. Over all n, the mean deviation equals 4.70 %.

The mean deviation of the production-rate estimate for n^* is a lot lower for the 20-station systems than for the 5- and 10-station configurations. The mean deviation over all n is higher than for n^* in both test sets. The better performance for 20-station systems for n^* is attributed firstly to the fact that the work-in-process estimation works very well for a medium utilization. Second, it is attributed (as we suppose) to the fact that the estimation error is compensated for by an increasing number of stations.[9] The worse performance over all n for 20-station systems is due to a more frequent non-convergence at a high number of n.

---

[9] This has been observed by Whitt (1984) for queueing models as well.

Chapter 5
Markov-Chain Approach

In this chapter, an analytical approach is presented for deriving the exact performance measures of closed queueing networks with finite buffer capacities, phase-type distributed processing times, and single servers. The closed queueing networks are represented by continuous-time Markov chains with discrete state space. We obtain the performance measures by setting up and solving a single Markov chain for the entire queueing system. Blocking is considered in blocking states that are subject to the phase-type distribution. The implementation allows an automated evaluation of an arbitrary number of configurations.

5.1 Overview of the Markov-Chain Approach

The basic concept of the approach is to expand the original configuration into a phase-type representation, analyze it, and aggregate it again. Figure 5.1 points out how the original and the phase-type representation of the CQN cohere. The figure further provides references to the sections which examine the particular issues.

If the processing times are not originally phase-type distributed, a suitable phase-type distribution is fitted from the processing rate μ_i and the coefficient of variation c_i at station i. The phase-type distribution is specified by the number of phases, ph_i^{NR}, by the rates of each phase ph_i at each station i, μ_{i,ph_i} $\forall i$, ph_i, and by the exit probabilities a_i at each station $i = 1, \ldots, M$.

The CQN is expanded by setting up the states space in the phase-type representation. The state space is determined by the phase-type distribution and the configuration of the CQN. A state is composed of the queue lengths $\overrightarrow{q}$, the current active phases $\overrightarrow{ph}$, and the blocking statuses, $\overrightarrow{bs}$, of each station. The transitions from the origin state s to the target state t, induced by station i are defined for all states s, t in the state set $\mathcal{S}$. The transition rates are denoted by μ_{sti}^{τ}. The states and transitions compose the Markov chain. From the built-up Markov chain, a system of

S. Lagershausen, *Performance Analysis of Closed Queueing Networks*, Lecture Notes in Economics and Mathematical Systems 663, DOI 10.1007/978-3-642-32214-3_5, © Springer-Verlag Berlin Heidelberg 2013

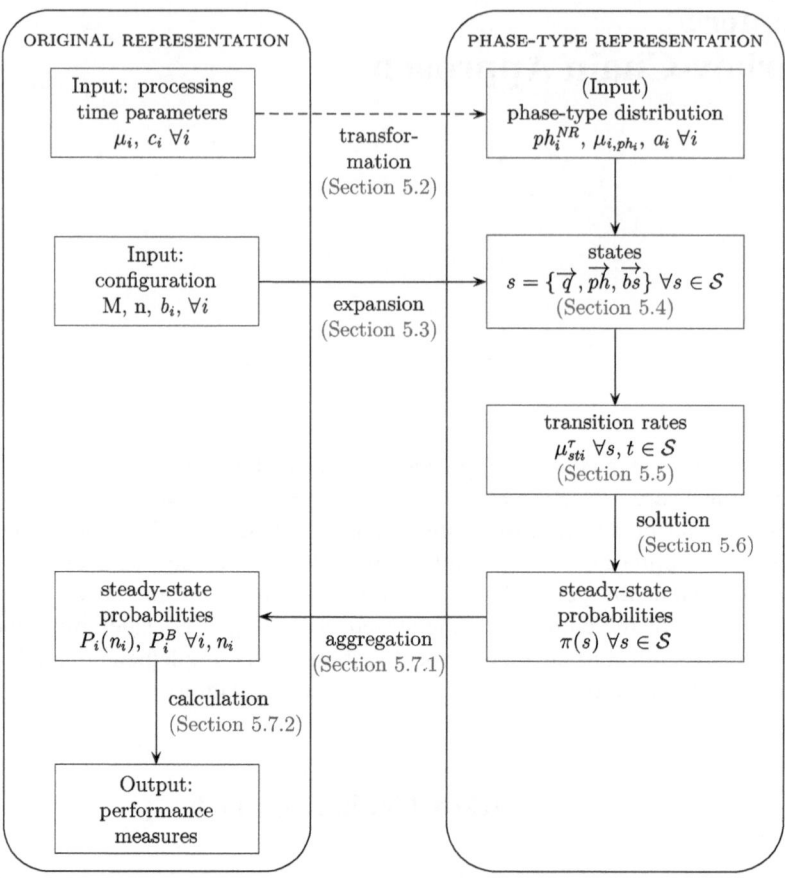

Fig. 5.1 Overview of the Markov-chain approach

linear equations is derived. The solution of these equations yields the steady-state probabilities $\pi(s)$ of each Markov-chain state $s \in \mathcal{S}$.

The steady-state probabilities of the original representation, $P_i(n_i) \; \forall i, n_i$ and the blocking probabilities $P_i^B \; \forall i$ are obtained by aggregating the $\pi(s)$. The $P_i(n_i)$ and P_i^B allow the calculation of the performance measures such as the production rate, the mean queue length, the cycle time and the utilization. The notation is given in Table 5.1.

This chapter is structured as follows. In Sect. 5.2, it is shown how processing time distributions specified by the first two moments are fitted to phase-type distributions with closed-form expressions. The selection of phase-type distributions regarding the minimization of the state space is explained in Sect. 5.2.1. According to the selection mechanism of that section, the parameter-fitting of selected phase-type distributions is outlined in Sect. 5.2.2.

Table 5.1 Notation

a_i	Probability of continuing the phase-type distribution at station i after the first phase
b_i	Buffer capacity at station i
$\overrightarrow{bs}$	Binary vector indicating which stations are blocked with $bs_i = 0$ if station i is not blocked and $bs_i = 1$ otherwise
c_i	Coefficient of variation of the processing time at station i
i	Station index, $i = 1, \ldots, M$
M	Number of stations
μ_i	Processing rate at station i, $\mu_i = \frac{1}{E[T_i]}$
μ_{i,ph_i}	Exponential processing rate of phase ph_i at station i
μ_{sti}^{τ}	Transition rate from state s to state t induced by station i
n	Number of workpieces in the system
n_i	Number of workpieces at station i
$\pi(s)$	Steady-state probability of state s
$P_i(n_i)$	Steady-state probability of n_i workpieces at station i
P_i^B	Probability of blocking at station i
$\overrightarrow{ph}$	Vector of active-phase indices over all stations
ph_i	Current active phase at station i
ph_i^{NR}	Number of phases of the phase-type distribution at station i
$\overrightarrow{q}$	Queue lengths at all stations, $q_i = \max\{n_i - 1, 0\}$
s	State
$\mathcal{S}$	Set of states in the Markov chain
T_i	Processing time at station i

In Sect. 5.3, the modeling of CQN by Markov chains is presented. First, in Sect. 5.3.1, the states and transitions are introduced. Next, in Sect. 5.3.2, examples of Markov-chain models are presented. The aspects of modeling are introduced step by step: The basic example considers exponential processing times and no blocking, see Sect. 5.3.2.1. The following example introduces the modeling of phase-type distributions, yet with infinite buffer capacities, see Sect. 5.3.2.2. Hereafter, blocking is introduced, first for the exponential distribution, see Sect. 5.3.2.3, and subsequently for the hypo-exponential-2 and the Cox-2 distribution, see Sect. 5.3.2.4. The three examples with blocking are traced throughout the succeeding sections.

The determination of the state space is presented in Sect. 5.4. First, all possible workpiece allocations are generated, see Sect. 5.4.1. Based on those, all combinations of active phases are created, see Sect. 5.4.2. Last, all possible blocking statuses are assigned, see Sect. 5.4.3.

Given all states, the relations between all states, the so-called transition rates, are obtained as shown in Sect. 5.5. In Sect. 5.5.1, the theoretical background is explained. Sect. 5.5.1.1 shows the assumptions and requirements underlying the Markov-chain model. These are then used in Sect. 5.5.1.2 to prove the global balance equations. The examples are continued with respect to the global balance equations in Sect. 5.5.1.3. The algorithm for the construction of the transition rate matrix is described in Sect. 5.5.2. The transition rate matrix is built up in two steps.

In the first step, the target states are specified for each possible transition, see Sect. 5.5.2.1. In the second step, the transition rate matrix is filled, see Sect. 5.5.2.2. An exemplary transition rate matrix is given in Sect. 5.5.2.3.

After the transition rate matrix has been set up, the corresponding system of linear equations resulting from the Markov chain can be solved. Its solution consists in the steady-state probabilities of the Markov chain. In Sect. 5.6, the solution method for the system of linear equations is presented, the so-called generalized minimal residual method (GMRES). It is a special method created to solve sparsely populated matrices. The results for the continued examples are provided in Sect. 5.6.2.

The calculation of the performance measures is presented in Sect. 5.7. For this purpose, first the steady-state probabilities of the Markov-chain states are aggregated, see Sect. 5.7.1. Subsequently, the performance measures are calculated from the steady-state probabilities, see Sect. 5.7.2. Examples are presented in Sect. 5.7.3. Lastly, in Sect. 5.8, the runtime performance is investigated, and numerical results are provided.

5.2   Phase-Type Distributions

Most procedures consider exponential service times and are, therefore, limited to a coefficient of variation c^2 equal to 1. The reason for the extensive use of the exponential distribution is its Markov property. This property (applied to processing times) is the reason that the residual processing time is distributed as the entire service time is.[1] That residual processing time is needed in numerous performance-analysis procedures, thus making the exponential distribution popular as an assumption. In the proposed Markov-chain approach, we use phase-type distributions. Since phase-type distributions are composed of several exponential phases, we benefit from the Markov property while still being able to apply any coefficient of variation. That is, however, at the cost of a higher computation time.

If the service time is not originally phase-type distributed, such a distribution can be fitted. For a given mean and variance of any value, closed-form expressions are available for the parameters of the phase-type distributions. Principally, the variability is reduced by connecting exponential phases in series because then, the process is stabilized. The variability is increased by routing the jobs with certain probabilities to exponential phases of different rates. This section shows how phase-type distributions are modeled to fit the service time parameters specified by the first two moments.

The concept of connecting exponential phases was introduced by Erlang (1917). The Erlang distribution holds for coefficients of variation $c^2 = \lceil \frac{1}{k} \rceil$ with $k \in \mathbb{N}$, where k represents the number of phases. Later, Cox (1955) generalized this concept

---

[1] See Sect. 5.2.2.

for any coefficient of variation. He showed that any distribution having a rational Laplace-Stieltjes transform can be represented by a set of exponential phases.[2] There are several other phase-type distributions. For an overview, see e.g. Stewart (2009).

Various moment-matching approximations and closed-form expressions have been proposed. Marie (1980) provided formulas to fit general distributions with a coefficient of variation of $c^2 > 0.5$ to phase-type distributions. Altiok (1985) presented closed-form expressions for the approximation of the first three moments of general distributions to phase-type distributions focusing on $c^2 > 1$. Johnson and Taaffe (1989) as well as Johnson and Taaffe (1991b) provided closed-form expressions for phase-type distributions matching the first three moments of a mixed Erlang distribution. However, this way, a high number of phases is required. In their consecutive work, Johnson and Taaffe (1990a, 1990b) assigned the hyper-exponential distribution with balanced means, where the first two moments are given, and fitted a mixture of two Erlang distributions for the first three moments. Their solution requires nearly the minimum number of phases; however, the algorithm has to solve a nonlinear problem. Asmussen, Nerman, and Olsson (1996) used the so-called EM Algorithm for a parameter fit of the first three moments. Osogami and Harchol-Balter (2006) suggested closed-form expressions for the first three moments. Johnson and Taaffe (1991a) showed that the third and higher-order moments do influence the results of the performance measures in a queueing system.

The approach we propose may incorporate any phase-type distribution and, therefore, an arbitrary number of moments can be considered. However, the allowance of more moments requires more phases, so one has to balance between the approximation quality and the computation time. In practice, usually only the mean, and at most additionally the variance is given. Hence, we focus on the first two moments for the parameter-fitting.

In Sect. 5.2.1, the selection of the phase-type distributions is described. In Sect. 5.2.2, the chosen phase-type distributions and the calculation of its parameters are presented.

5.2.1 Selection of Phase-Type Distributions

Each phase-type distribution can model any mean value, but only a limited range of coefficients of variation. Conversely, for each coefficient of variation, several phase-type distributions exist. Thus, a phase-type distribution must be chosen according to the given coefficient of variation. The first selection criterion for the Markov-chain approach is the proper fit of μ and c^2. The second criterion constitutes the minimum number of phases and the third embodies the minimum number of transitions. This

---

[2]See Altiok (1996, p. 57).

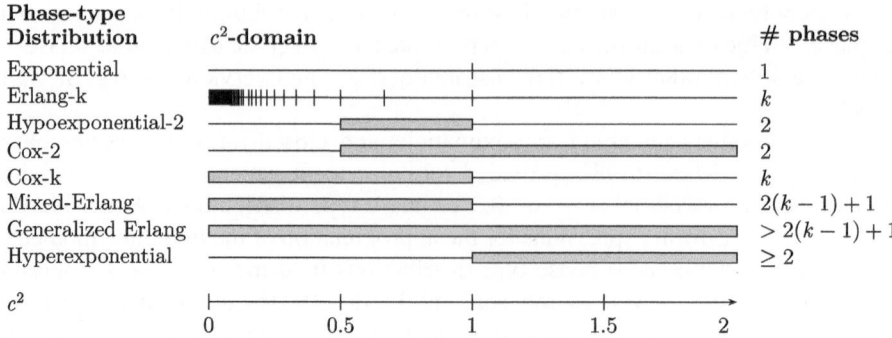

Fig. 5.2 Applicability of phase-type distributions depending on c^2

selection procedure ensures the minimum computation time and internal memory requirement for given μ and c^2.

Figure 5.2 depicts the range of c^2 for the most common phase-type distributions.[3] The number of required phases for each distribution is listed in the right-hand column. These result from the closed-form expressions of the two-moment parameter fits as presented in Bolch, Greiner, de Meer, and Trivedi (2006) and Tijms (2003). In Fig. 5.2 holds that $k = \lceil 1/c^2 \rceil$. For $c^2 = 1$, the exponential distribution has to be used. In the case of $c^2 > 1$, the Cox-2, the generalized Erlang and the hyper-exponential distribution are valid alternatives. The generalized Erlang distribution, though being able to model the complete range of c, requires at least $2(k-1)+1$ phases. The hyper-exponential distribution needs at least two phases, whereas for the Cox-2 distribution, two phases suffice. Therefore, the Cox-2 distribution is the best choice for $c^2 > 1$ under the aforementioned objectives.

In the domain of $0.5 \leq c^2 < 1$, both the hypo-exponential and the Cox-2 distribution require the least number of phases, namely 2. In the hypo-exponential distribution, one less transition is carried out because branching does not occur. Thus, the transition rate matrix contains fewer entries which leads to a shorter solution time. Hence, the hypo-exponential distribution is chosen.

In the case of $c^2 \leq 0.5$, the Erlang-k, the Cox-k, the mixed-Erlang and the generalized Erlang distribution are potential distribution models. The number of phases is minimized by applying the Erlang-k and the Cox-k distribution. For a decision between those two, the number of transitions is consulted as a second criterion. As a result, the Erlang-k distribution is applied if $k = 1/c^2 \in \mathbb{N}$, and the Cox-k distribution is used otherwise.

Table 5.2 summarizes the assignment of the phase-type distributions subject to a given c^2. The displayed association corresponds to the selection of phase-type distributions in this Markov-chain approach.

---

[3]See Rall (1998, p. 115).

Table 5.2 Assignment of phase-type distributions for different c^2

	c^2		Distribution	# phases
	c^2	> 1	Cox-2-distribution	2
	c^2	$= 1$	Exponential distribution	1
$0.5 <$	c^2	< 1	Hypo-exponential distribution-2	2
	c^2	$= \frac{1}{k}$	Erlang-k distribution	k
	c^2	< 0.5	Cox-k distribution	k

Table 5.3 Notation

a	Probability of continuing the phase-type distribution after the first phase
c	Coefficient of variation of the processing time
$E(T)$	Expected processing time
μ	Processing rate at a station; reciprocal of the expected processing time: $\mu = 1/E(T)$
μ_p	Exponential processing rate of the p-th phase
k	Number of phases in the phase-type distribution
T	Random variable representing the processing time

Fig. 5.3 Exponential distribution

5.2.2   *Two-Moment Parameter Fit of Selected Phase-Type Distributions*

In the following, the two-moment parameter fit for the selected phase-type distributions is presented. Based on the moment-generating function of the phase-type distribution, the formulas for the expected value, $E(T)$, and the coefficient of variation, $c^2 = \sqrt{\frac{Var(T)}{E(T)^2}}$, are provided. A rearrangement of these formulas yields the expressions for the parameters of the distribution subject to μ and c^2. If a phase-type distribution has more than two parameters, additional assumptions have to be made (because then only two equations, but more unknown parameters exist). The assumptions are introduced where the distributions are presented. The notation of this section is given in Table 5.3.

Exponential distribution. The exponential distribution represents the basis of the phase-type distributions. It consists of one exponential phase, which has the rate μ. A graphical representation is depicted in Fig. 5.3.

The parameter μ of the exponential distribution corresponds to the reciprocal of the expected value $E(T)$, see Eq. (5.1). The variance is given by $Var(T) = \mu^2$, and the coefficient of variation equals 1, $c = 1$.

$$\mu = \frac{1}{E(T)} \tag{5.1}$$

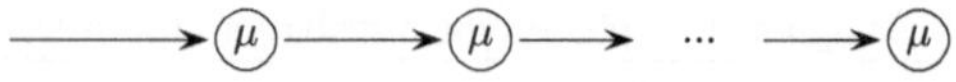

Fig. 5.4 Erlang-k distribution

$$\longrightarrow (\mu_1) \longrightarrow (\mu_2) \longrightarrow$$

Fig. 5.5 Hypo-exponential-2 distribution

Erlang-k distribution. The Erlang-k distribution, proposed by Erlang (1917), consists of k phases, each with the same exponential rate μ. The phases are connected in series as shown in Fig. 5.4. The expected value and the coefficient of variation (CV) are given in Eqs. (5.2) and (5.3).

$$E(T) = \sum_{j=1}^{k} \frac{1}{\mu} = \frac{k}{\mu} \qquad (5.2)$$

$$c = \frac{1}{\sqrt{k}} \qquad (5.3)$$

The parameters k and μ are obtained by a rearrangement of (5.2) and (5.3), see Eqs. (5.4) and (5.5). The Erlang-k distribution requires the term $1/c^2$ to be an integer value.[4]

$$k = \frac{1}{c^2} \qquad \qquad \text{if } \frac{1}{c^2} \in \mathbb{N} \qquad (5.4)$$

$$\mu = \frac{k}{E(T)} \qquad (5.5)$$

Hypo-exponential-2 distribution. The hypo-exponential-2 distribution is characterized by two succeeding exponential phases with different rates μ_1 and μ_2 with $\mu_1 \neq \mu_2$. It is depicted in Fig. 5.5. The expected value and the coefficient of variation are given in (5.6) and (5.7).

$$E(T) = \frac{1}{\mu_1} + \frac{1}{\mu_2} \qquad (5.6)$$

$$c = \frac{\sqrt{\mu_1^2 + \mu_2^2}}{\mu_1 + \mu_2} < 1 \qquad (5.7)$$

---

[4]If k is a real non-integer number, a gamma distribution is modeled. The gamma distribution represents a generalization of the Erlang-k distribution. Since $k \in \mathbb{R}^+$ cannot represent the number of phases, the gamma distribution is not a phase-type distribution.

Fig. 5.6 Cox-2 distribution

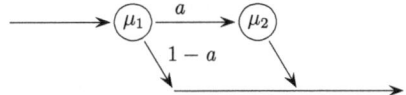

Solving the equations for μ_1 and μ_2, the parameters of the distribution are obtained. The expressions are displayed in Eqs. (5.8) and (5.9).

$$\mu_1 = \frac{2}{E(T)} \cdot \left[1 + \sqrt{1 + 2 \cdot (c^2 - 1)}\right]^{-1} \tag{5.8}$$

$$\mu_2 = \frac{2}{E(T)} \cdot \left[1 - \sqrt{1 + 2 \cdot (c^2 - 1)}\right]^{-1} \tag{5.9}$$

If $\mu_1 = \mu_2$, an Erlang-2 distribution with $c^2 = 0.5$ is modeled. Concurrently, $c^2 = 0.5$ represents the lower bound of c^2 that can be modeled by the two exponential phases connected in series. The upper bound for c^2 that can be modeled by the hypo-exponential distribution is smaller than 1, see Eq. (5.7).[5]

Cox-2 distribution. The Cox-2 distribution consists of two exponential phases. The first phase with rate μ_1 is always completed. With probability a, also the second phase with rate μ_2 is accomplished. With the counter-probability $1 - a$, the process is completed after the first phase. This splitting adds variability to the process. The Cox-2 distribution is used here to model coefficients of variation greater than 1, $c > 1$. A graphic of the Cox-2 distribution is depicted in Fig. 5.6. The expected value and the coefficient of variation are given in Eqs. (5.10) and (5.11).

$$E(T) = \frac{1}{\mu_1} + \frac{a}{\mu_2} \tag{5.10}$$

$$c = \frac{\mu_2^2 + a\mu_1^2(2 - a)}{(\mu_2 + a\mu_1)^2} \tag{5.11}$$

For this distribution, an infinite number of solutions exists, as there are two equations for the three parameters μ_1, μ_2, and a. Therefore, an assumption is made: the expected residence time in the first phase equals half of the expected service time: $1/\mu_1 = E(T)/2$. With this commitment, the parameters a and μ_2 are fitted to meet the desired c^2 and $E(T)$ according to (5.10) and (5.11). The resulting expressions of the parameters are shown in Eqs. (5.12)–(5.14).[6]

$$\mu_1 = \frac{2}{E(T)} \tag{5.12}$$

---

[5] See Bolch et al. (2006, p. 24).
[6] See Bolch et al. (2006, p. 28).

Fig. 5.7 Cox-k distribution

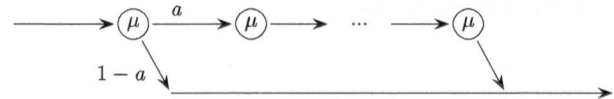

$$\mu_2 = \frac{1}{E(T) \cdot c^2} \qquad (5.13)$$

$$a = \frac{1}{2 \cdot c^2} \qquad (5.14)$$

Cox-k distribution. The Cox-k distribution consists of k phases of equal rate μ connected in series as in the Erlang-k distribution. Furthermore, there is a branching after the first phase, as in any Cox-distribution. For the Cox-k distribution it holds that $c^2 < 1$. The arrangement goes back to Sauer and Chandy (1975). In Rall (1998) and Marie (1980), this distribution model is called generalized Erlang. However, in Bolch et al. (2006), it is named Cox-distribution.[7] The scheme of the distribution, named here as Cox-k distribution, is depicted in Fig. 5.7.[8]

The expected value and the CV are given in Eqs. (5.15) and (5.16).

$$E(T) = \frac{(1-a) + k(1-(1-a))}{\mu} \qquad (5.15)$$

$$c = \frac{k + (1-a)(k-1)((1-a)(1-k) + k - 2)}{[(1-a) + k(1-(1-a))]^2} \qquad (5.16)$$

Using the two equations (5.15) and (5.16) for the three parameters a, μ and k, again an infinite number of solutions exists. This is settled by setting k to the minimum number of phases. The number of required phases amounts to $1/c^2$, by which k is specified, see Eq. (5.17). The other parameters result from this assumption and are given in Eqs. (5.18)–(5.19).

$$k = \left\lceil \frac{1}{c^2} \right\rceil \qquad (5.17)$$

$$a = 1 - \frac{2kc^2 + (k-2) - \sqrt{k^2 + 4 - 4kc^2}}{2(k-1)(c^2 + 1)} \qquad (5.18)$$

$$\mu = \frac{(1-a) + a \cdot k}{E(T)} \qquad \text{with } 0 \le b_i \le 1 \qquad (5.19)$$

The Cox-k distribution equals the Erlang distribution if $1/c^2 \in \mathbb{N}$ and $a = 1$.

---

[7] See Bolch et al. (2006, p. 27).

[8] The general Cox-k distribution has branches leading out of the process after each phase, cf. e.g. Bolch et al. (2006, p. 27f). For the Cox-k distribution, it is assumed that $a_i = 1 \ \forall i > 1$.

5.3 Modeling

In the following, we show how CQN are modeled by Markov chains. In Sect. 5.3.1, exemplary states and transitions are presented in order to emphasize the expansion of the phase-type representation. In Sect. 5.3.2, complete Markov chains are introduced.

5.3.1 Phase-Type and Original Representation

Let us consider the following configuration of a CQN: The number of stations equals $M = 3$, the service time distribution at station 1 follows a hypo-exponential-2 distribution, the service time distribution of station 2 is a Cox-2 distribution, and that of station 3 is supposed to be an Erlang-3 distribution. Figure 5.8 depicts this CQN with the corresponding phase-type distributions. The taller rectangles represent the service stations and the smaller rectangles each embody one unit of buffer space.

The conglomerate of exponential phases represents the model of the service time distribution. The processing time is given by the complete passage of phases. Consequently, only a single job can reside in the arrangement of phases at any instant in time. The job that is next to be processed can enter the service station as soon as the preceding job exits the last phase.

A possible sequence of states with $n = 6$ customers is depicted in Fig. 5.9. The depicted sequence of states represents one of many possible sequences. Each diagram of the CQN represents a state of the system. The filled circles indicate the positions of the workpieces, whereas the workpiece in focus is highlighted by the color red. In diagram (a), two workpieces are located in buffer 1, and one workpiece is situated in buffer 3. Station 1 processes in the first phase, station 2 is active in the second phase, and at station 3, the job-in-service resides in the first phase. For a change of this state, one of the currently active phases must be finished.

A departure from state (a) occurs if the current active phase at station 3 is finished first out of all the active phases. If this happens, the system transits into the state depicted in diagram (b). In this state, the red-colored job resides in the second phase. Assuming that in state (b) the second phase of station 3 is completed next, the job moves to the third phase which results in the state of diagram (c). With the red-colored job being finished first once again, the last phase is completed, and the job is forwarded to the succeeding station, see diagram (d).

The phase-type representation of states serves as a model of the CQN-processes in the Markov-chain approach. In the original representation, a station can only be active, starving or blocked; no further distinctions are made. In the phase-type representation, however, the single manufacturing process is disaggregated into several exponential phases. This implies an expansion of states, but enables the exact

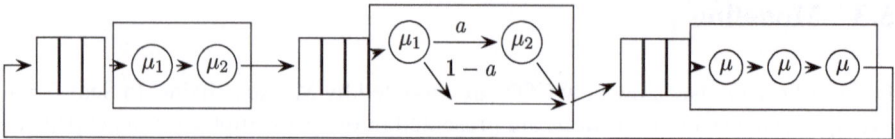

Fig. 5.8 CQN with phase-type distributions

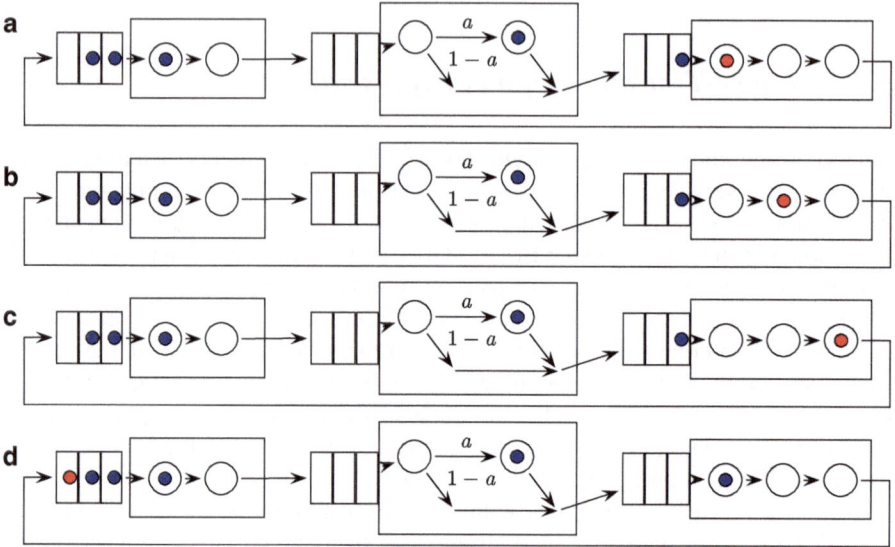

Fig. 5.9 Exemplary transition of customers in the phase-type representation

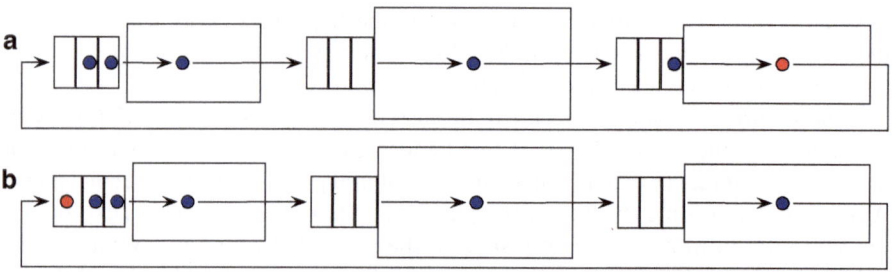

Fig. 5.10 Corresponding states in the original representation

analysis (provided the service time follows a phase-type distribution). Figure 5.10 depicts the corresponding states in the original representation. Herein, the states (a), (b), and (c) of the phase-type representation correspond to one state in the original representation.

Fig. 5.11 Two-station system

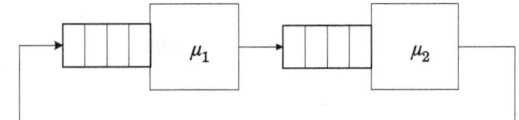

Fig. 5.12 Markov chain of a two-station system with $n = 3$

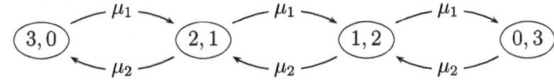

5.3.2 Markov-Chain Odels of CQN

The Markov chain of a CQN is composed of all states and all transitions between those states. In the following, we present examples of Markov-chain models for CQN with exponential and phase-type distributions as well as with infinite and finite buffer capacities.

5.3.2.1 Exponential Distribution Without Blocking

In this section, a CQN with an exponential processing time distribution and infinite buffer capacity is treated. An example with $M = 2$ stations is depicted in Fig. 5.11. Further assuming $n = 3$ customers, the Markov chain displayed in Fig. 5.12 results.

In this setting, a state is represented by the number of customers n_i over the stations $i = 1, 2$. Each state is denoted by (n_1, n_2). The first workpiece at a station is assumed to receive service. Depending on the state, specific transitions into other states are possible. For example, state $(n_1, n_2) = (3, 0)$ can only be exited when station 1 finishes its current job. If this happens, the system transits into state $(2, 1)$ with the rate μ_1. From state $(2, 1)$, two different transitions are possible: If station 1 finishes first, the system changes into state $(1, 2)$ with rate μ_1. Otherwise, if station 2 finishes first, the system transits into state $(3, 0)$ with rate μ_2.[9] From state $(1, 2)$, the system either transits into state $(0, 3)$ with rate μ_1 or it reaches state $(2, 1)$ with rate μ_2. State $(0, 3)$ can only be exited with rate μ_2.[10]

5.3.2.2 Phase-Type Distribution Without Blocking

The analysis of CQN with phase-type distributed processing times works analogously. Let us again consider $n = 3$ customers in a two-station system with infinite buffer capacities. Figure 5.13 shows the two-station queueing system. The

---

[9]Note, that this implies the Markov-property. For further details, see Sect. 5.5.1.

[10]See Kleinrock (1975, pp. 156ff).

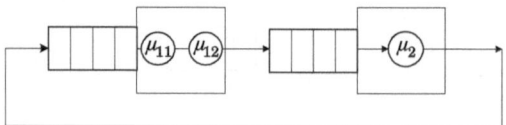

Fig. 5.13 Two-station system with hypo-exponential-2 distribution

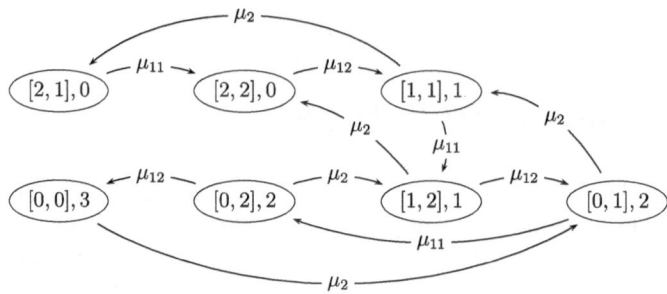

Fig. 5.14 Markov chain of a two-station system with hypo-exponential-2 distribution

processing time at station 1 is assumed to be hypo-exponential-2 distributed and the processing time at station 2 follows an exponential distribution. The hypo-exponential-2 distribution consists of two successive exponential phases.[11] The rates are denoted by μ_{i,ph_i} with ph_i representing the index of the phase at station i. The states of the corresponding Markov chain, which is depicted in Fig. 5.14, are denoted by $([q_1, ph_1], n_2)$. q_1 indicates the queue length at station 1, ph_1 represents the index of the active phase at station 1, and n_2 denotes the number of customers at station 2.[12] The notation is summarized in Table 5.4.

When the system is in state $([q_1, ph_1], n_2) = ([2, 1], 0)$, a change of states can only arise if phase 1 at station 1 is finished. The transition into state $([2, 2], 0)$ occurs with rate $\mu_{i,ph_i} = \mu_{11}$. If the second and last phase of processing is finished at station 1 in state $([2, 2], 0)$, the customer is transferred to the next station. Then, at station 1, the first customer in the queue enters the first phase and the queue length q_1 is reduced by 1. Further, station 2 starts the processing of the transferred customer. The resulting state, $([1, 1], 1)$, is entered from state $([2, 2], 0)$ with rate μ_{12}. In state $([1, 1], 1)$, there are two possible state changes. One possibility is that station 2 finishes its customer first. In this case, the customer is led into the buffer in front of station 1 and the system transits into state $([2, 1], 0)$ with rate μ_2. Alternatively, phase 1 at station 1 is finished first. In this instance, the customer at station 1 is transferred into the second processing phase. All other transitions take place accordingly.

---

[11]For more information on the hypo-exponential-2 distribution, see Sect. 5.2.2 on page 68.
[12]See Bolch et al. (2006, pp. 332ff).

Table 5.4 Notation

μ_{i,ph_i}	Exponential rate at station i in phase ph_i
n_i	Number of workpieces at station i (waiting, in processing, or blocked), with $n_i = 0, \ldots, \min\{n, b_i + 1\}$
$\pi(s)$	Steady-state probability of residing in state s
ph_i	Index of active phase at station i, $ph_i = \mathbb{1}_{\{n_i \geq 1\}}$
$P_i(n_i)$	Steady-state probability of n_i workpieces at station i
P_i^B	Blocking probability at station i
q_i	Queue length at station i, $q_i = \max\{n_i - 1, 0\}$
s	Index of a state in the Markov chain
$\mathcal{S}$	State space of the Markov chain

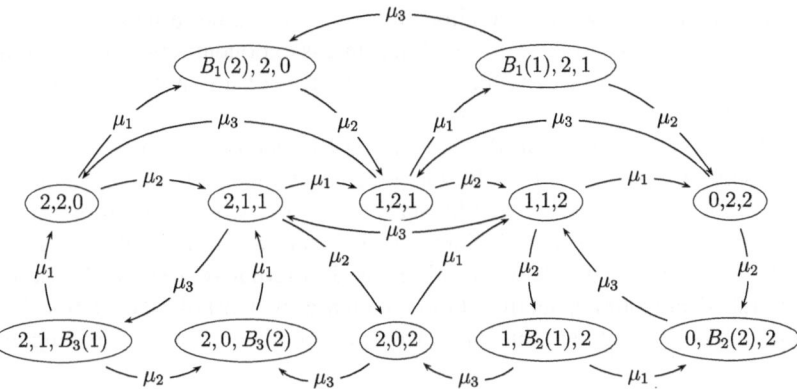

Fig. 5.15 Markov chain of a three-station system with exponential distribution and blocking (**Example 1**)

5.3.2.3 Exponential Distribution with Blocking

In the following example, the processing times are exponentially distributed at all stations; however, the buffer space is limited. In this setting, a state is described by the number of workpieces at each station and the blocking status of the server. If no blocking occurs, only the workpiece allocation is annotated. For example, the notation $(2, 2, 0)$ describes the state in which there is one workpiece in the queue and one in service, both at station 1 and 2 while station 3 is starving. If station i is blocked and n_i workpieces are located at station i, the notation reads $B_i(n_i)$.

Figure 5.15 depicts the Markov chain of a three-station closed queueing network with $n = 4$ customers circulating in the system. The service times are exponentially distributed with the rate μ_i at the stations $i = 1, 2, 3$. There is one unit of buffer space in front of each station i, i.e. $b_i = 1$ $\forall i$. This example will be continued throughout the chapter and is referred to as **Example 1**.

Let us consider the departures from state $(2, 2, 0)$. This state is left by either station 1 or station 2 finishing its current job. If station 2 is first to finish the job-in-service, the system transits into state $(2, 1, 1)$. This happens with the processing rate of station 2, μ_2. The other possibility of leaving state $(2, 2, 0)$ is for station 1 to finish its job first. If this occurs, station 1 becomes blocked. The workpiece allocation stays the same, but the blocking status changes. Hence, state $(B_1(2), 2, 0)$ is entered.

There are two reasons for the establishment of a new state in this case. First, the blocking probability is needed for the performance evaluation. It can only be computed if the blocking states are taken into account separately. Second, the transition rates of this state are different from its non-blocked equivalent. Regarding our example, this means: In contrast to state $(2, 2, 0)$, from state $(B_1(2), 2, 0)$, the only way out is for station 2 to finish its current job. The different transition rates change the relative residence times and, therefore, the steady-state probabilities. As a result, although the workpiece allocation stays the same, a new state has to be entered. The residence time of the blocking state is exponentially distributed (in this case with parameter μ_2), as required for each state of a continuous-time Markov chain.

Another important issue in the modeling of queueing networks with blocking is the communication between the blocking states. Consider, for example, state $(B_1(1), 2, 1)$: Station 1 is blocked by station 2 with a workpiece on its server and none in the buffer. One possible change is that station 2 finishes its job. Then, the blocking is resolved and state $(0, 2, 2)$ is entered. Alternatively, station 3 could finish its job first. That would mean that station 3 releases its workpiece to station 1, while station 1 stays blocked, with one more workpiece in the buffer, thus leading to state $(B_1(2), 2, 0)$.

5.3.2.4   Phase-Type Distributions with Blocking

In the following, two Markov-chain examples for CQN with phase-type distributions and blocking are presented. In the first example, the hypo-exponential-2 distribution is considered and in the second example, the Cox-2 distribution.

Hypo-exponential-2 distribution with blocking. We consider Example 1, except that the processing time at station 1 follows a hypo-exponential-2 distribution. The resulting Markov chain is depicted in Fig. 5.16. The states are denoted by $([q_1, ph_1], n_2, n_3)$. q_1 indicates the queue length of the buffer in front of station 1, and ph_1 specifies the index of the active phase at station 1. n_2 and n_3 denote the number of workpieces at stations 2 and 3. The rate of the first phase is labeled by μ_{11}, and the rate of the second phase is indicated by μ_{12}. This example is also continued and referred to as **Example 2**.

Consider state $([q_1, ph_1], n_2, n_3) = ([1, 1], 2, 0)$. In this state, at station 1, one workpiece resides in the buffer, and one workpiece receives service in the first phase. At station 2, also one workpiece waits in the buffer, and one is located on the server. There are two possible departures from this state. One possibility is that station 2 completes its current job. Then, this job is passed on to station 3 and the system

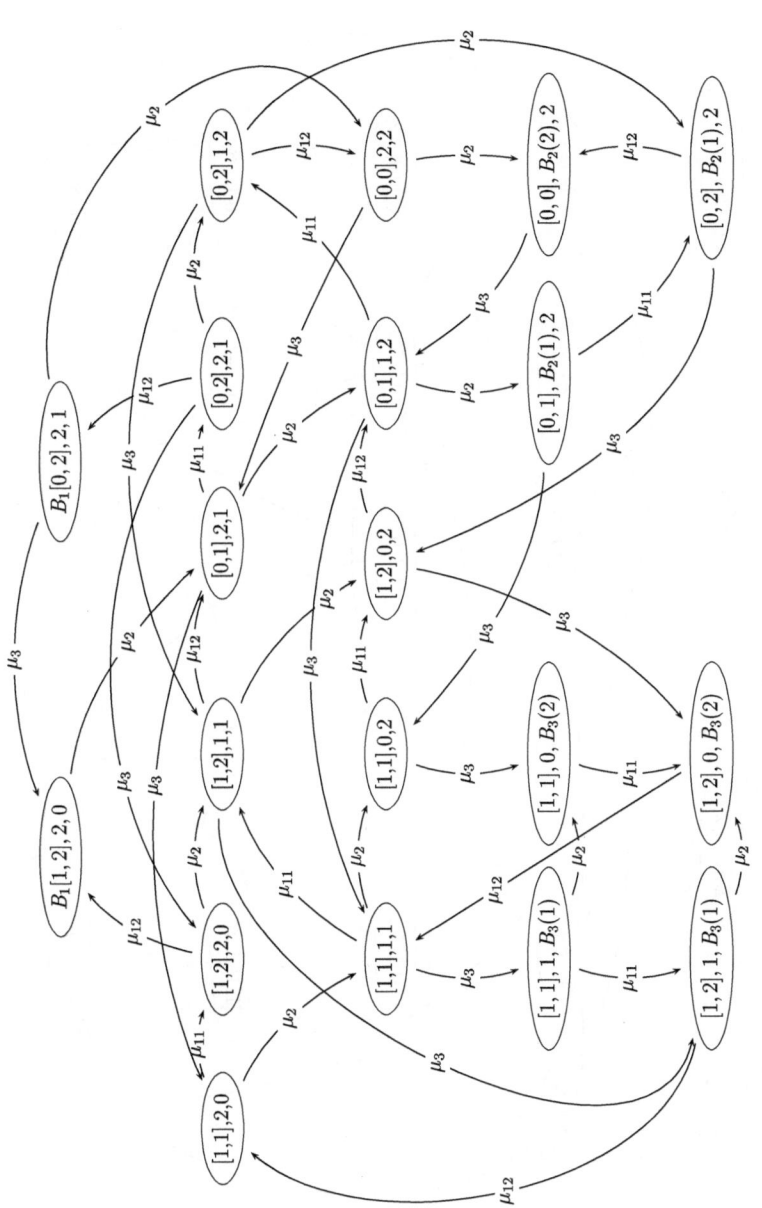

Fig. 5.16 Markov chain with three stations and a hypo-exponential-2 distribution (**Example 2**)

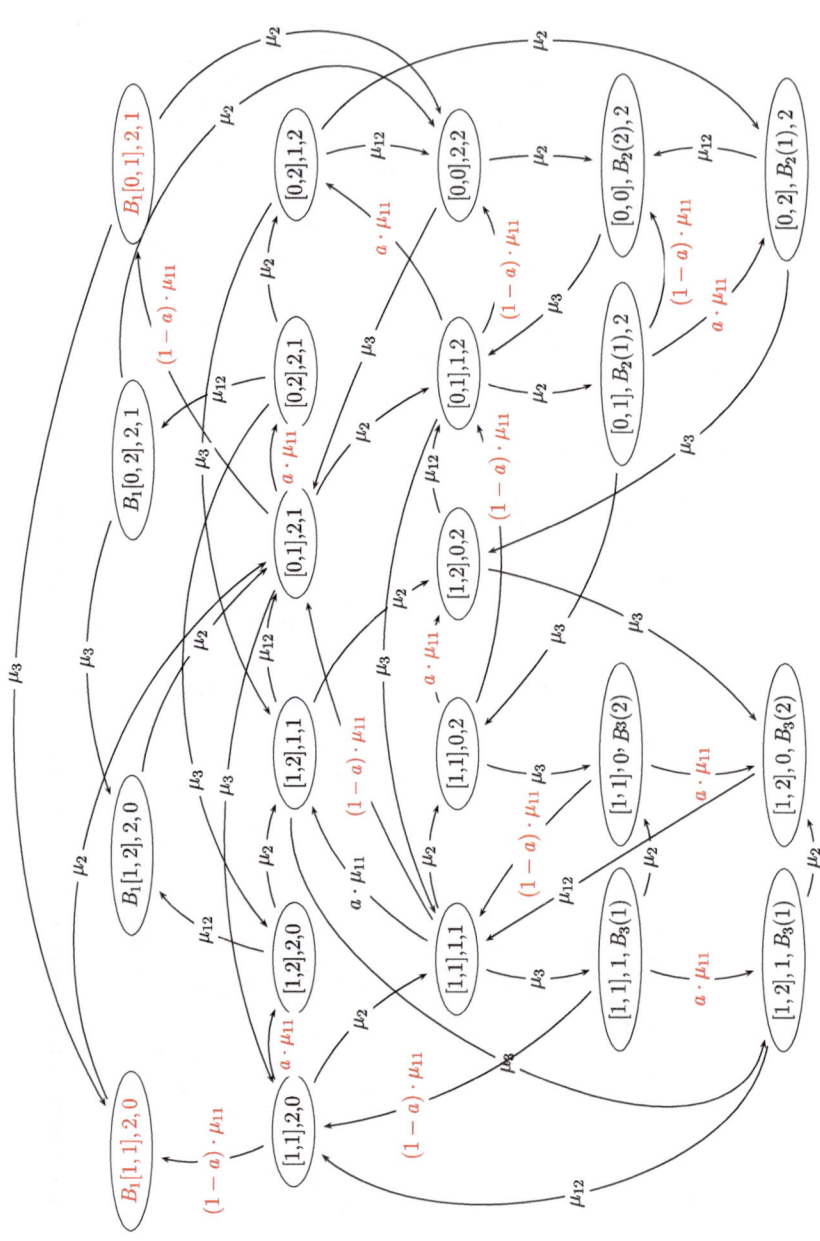

Fig. 5.17 Markov chain with three stations and a Cox-2 distribution at station 1 (**Example 3**)

transits into state ([1,1],1,1) with rate μ_2. The other possibility is that station 1
finishes the current phase. In this case, the system transits into state ([1,2],2,0) with
rate μ_{11}.

The necessary condition for station i to become blocked (regardless of the
processing time distribution) constitutes that the buffer between station i and $i + 1$
is full, which means $q_{i+1} = b_{i+1}$. In the case of phase-type distributed processing
times, the additional condition that needs to be fulfilled reads that station i must be
active in the last processing phase. In Example 2, station 1 becomes blocked, based
on residing in state ([1,2],2,0), if station 1 finishes the second phase before station 2
completes its current workpiece. Then no buffer space is available in the succeeding
buffer, and the system transits into state ($B_1[1, 2], 2, 0$) with rate μ_{12}.

Cox-2 distribution with blocking. The Cox-2 distribution consists of two exponential phases with an additional exit possibility after the first phase.[13] The job may
be finished after the first phase with probability $(1 - a)$. With probability a, the
service is pursued with the second phase. The branching in the Cox-2 distribution
leads to more transitions between the states and to more blocking states as compared
to the hypo-exponential-2 distribution.

The Markov chain of the three-station example with Cox-2 distributed processing
times at station 1 is depicted in Fig. 5.17. The changes in comparison to the hypo-
exponential-2 distribution are highlighted in red. The example is referred to as
Example 3 throughout this chapter.

Let us consider state ([1,1],2,0). With probability a, the processing of the
workpiece-in-service at station 1 needs two phases. If this is the case and if further
phase 1 at station 1 is finished before the processing at station 2, the system will
transit into state ([1,2],2,0) with rate $a \cdot \mu_{11}$. Otherwise, if phase 1 is still shorter
than the processing time at station 2, but the process at station 1 takes only one
phase (this occurs with probability $(1 - a)$), station 1 will become blocked and the
system will transit into state ($B_1[1, 1], 2, 0$) with rate $(1 - a) \cdot \mu_{11}$. When the system
resides in state ([1,2],2,0), station 1 becomes blocked with rate μ_{12}, resulting in state
($B_1[1, 2], 2, 0$) .

5.4 State Space

As stated above, the recurrence time of each state in a continuous-time Markov
chain must be exponentially distributed. By modeling the states in the phase-type
representation, this is ensured.[14] This section presents an algorithm to generate the
states of CQN in this phase-type representation. The set of states depends on the
given configuration of the CQN (described by the number of stations M, the number

---

[13]For a graphical representation of the Cox-2 distribution, see Fig. 5.6 on page 69.

[14]For more information, see Sect. 5.5.1.

Table 5.5   Notation

a_i	Probability of continuing with the next phase in the phase-type distribution at station i
$\overrightarrow{b}$	Buffer capacity over all stations
$\overrightarrow{bp}$	Vector indicating if a station can potentially be blocked, $bp_i = 0$ if station i cannot be blocked and $bp_i = 1$ if station i can be blocked
$\overrightarrow{bs}$	Binary vector indicating which stations are blocked with $bs_i = 0$ if station i is not blocked and $bs_i = 1$ otherwise
$\overrightarrow{bs}^I$	Initial blocking statuses, $bs_i^I = 0 \ \forall i$
i	Station index, $i = 1, \ldots, M$
M	Number of stations
n	Number of workpieces in the system
$\overrightarrow{ph}^I$	Initial active phases, $ph_i^I = 1$ if $w_i \geq 1$ and $ph_i^I = 0$ if $w_i = 0$
$\overrightarrow{ph}^{NR}$	Number of phases according to the phase-type distribution
$\overrightarrow{PTD}$	Phase-type distributions at all stations
$\overrightarrow{q}$	Queue lengths at all stations, $q_i = \max\{w_i - 1, 0\}$
s	State
$\mathcal{S}$	Set of states in the Markov chain
$\overrightarrow{w}$	Workpiece allocation over all stations

of customers n, and the buffer sizes $b_i \ \forall i$) and the phase-type distribution for all stations (specified by the number of phases of the distribution ph_i^{NR} at each station i, the exponential rates of the ph-th phase, denoted by $\mu_{i,ph_i} \ \forall i$, $ph_i = 1, \ldots, ph_i^{NR}$, and the exit probabilities at station i, indicated by $a_i \ \forall i$). A state s is fully determined by the queue length $\overrightarrow{q}$, the indices of the current active phases $\overrightarrow{ph}$, and the blocking statuses $\overrightarrow{bs}$, each for every station, see Eq. (5.20).

$$s = \{\overrightarrow{q}, \overrightarrow{ph}, \overrightarrow{bs}\} \tag{5.20}$$

The vector length of each vector corresponds to the number of stations, M. The state set of all states s in the phase-type distribution of the CQN is denoted by $\mathcal{S}$. The notation used in this section is given in Table 5.5.

Algorithm 7 provides an overview of the complete procedure for the generation of states. It is composed of three sub-procedures. Each time a procedure is called, it generates another realization of its corresponding part of the state (i.e. the workpiece allocations over the stations, the indices of the current active phases, or the blocking statuses). These sub-procedures are called repeatedly until all combinations are created.

For given M, n, and $\overrightarrow{b}$, a specific workpiece allocation $\overrightarrow{w}$ is generated in the sub-procedure WORKPIECE ALLOCATION. This procedure is described in Sect. 5.4.1. From the workpiece allocation, the queue lengths $q_i = \max\{w_i - 1, 0\}$ are derived. Together with initial phase-values $\overrightarrow{ph}^I$ and initial blocking statuses $\overrightarrow{bs}^I$, a state is composed and added to the state set.

Algorithm 7 Overview of procedure to generate states

1: **procedure** STATES $M, n, \overrightarrow{b}, \overrightarrow{PTD}$
2:   **repeat**
3:     Generate workpiece allocation $\overrightarrow{w}$ ▷ WORKPIECE ALLOCATION
4:     $\mathcal{S} := \mathcal{S} \cup s(\overrightarrow{q}, \overrightarrow{ph}^I, \overrightarrow{bs}^I)$
5:     Generate vector of upper bounds for phases $\overrightarrow{ph}^{UB}$
6:     **repeat**
7:       Generate a combination of active phases $\overrightarrow{ph}$ ▷ PHASES
8:       $\mathcal{S} := \mathcal{S} \cup s(\overrightarrow{q}, \overrightarrow{ph}, \overrightarrow{bs}^I)$
9:       **if** blocking is possible **then**
10:         Generate vector of potentially blocked stations $\overrightarrow{bp}$
11:         **repeat**
12:           Generate blocking statuses $\overrightarrow{bs}$ ▷ BLOCKING
13:           **for each** $\overrightarrow{bs}$ **do**
14:             $\mathcal{S} := \mathcal{S} \cup s(\overrightarrow{q}, \overrightarrow{ph}, \overrightarrow{bs})$
15:           **end for**
16:         **until** $\overrightarrow{bs} = \overrightarrow{bp}$
17:       **end if**
18:     **until** $\overrightarrow{ph} = \overrightarrow{ph}^{UB}$
19:   **until** all possible $\overrightarrow{w}$ are generated for given $M, n, \overrightarrow{b}, \overrightarrow{PTD}$
20: **end procedure**

In the next step, a vector of upper bounds for the active phases, $\overrightarrow{ph}^{UB}$, is derived from the given $\overrightarrow{w}$ and the specified phase-type distributions $\overrightarrow{PTD}$. With the vector $\overrightarrow{ph}^{UB}$ as input, a combination of active phases is generated in the procedure PHASES, see Sect. 5.4.2. Then, a state is composed with the given queue lengths $\overrightarrow{q}$, the generated indices of the active phases $\overrightarrow{ph}$ and initial blocking statuses $\overrightarrow{bs}^I$.

For the given values of $\overrightarrow{q}$ and $\overrightarrow{ph}$, it is then checked whether blocking can occur. If this is the case, a vector of potentially blocked stations is generated, denoted by $\overrightarrow{bp}$. Using $\overrightarrow{bp}$ in the procedure BLOCKING, all blocking statuses $\overrightarrow{bs}$ are derived, see Sect. 5.4.3. From each blocking-status combination $\overrightarrow{bs}$, together with the given $\overrightarrow{q}$ and the combination of the active phases $\overrightarrow{ph}$, another state is composed. If blocking cannot occur, the procedure BLOCKING is simply omitted.

After the determination of all blocking combinations for given $\overrightarrow{w}$ and $\overrightarrow{ph}$, the next combination of the active phases $\overrightarrow{ph}$ for a given workpiece allocation $\overrightarrow{w}$ are generated in the procedure PHASES. For each combination of active phases $\overrightarrow{ph}$, all blocking combinations are generated in the procedure BLOCKING, if blocking is possible. After the creation of all blocking statuses for a given $\overrightarrow{ph}$, the next $\overrightarrow{ph}$ is generated. After the generation of all combinations of the active phases and blocking statuses for a given workpiece allocation $\overrightarrow{w}$, the next workpiece allocation

Table 5.6 Notation

d_i	Upper bound of the number of workpieces at station i, $d_i = \min\{n, b_i + 1\}$
f	Highest station index for which the workpiece allocation is fixed (starting from the first station, i.e. $i = 1, \ldots, f$)
g_i	Station capacity, $g_i = b_i + 1$
i^{UB}	Lowest station index containing a workpiece $i^{UB} = \arg \min_i \{i = 1, \ldots, M \,|\, w_i > 0\}$
n^R	Number of remaining workpieces, $n^R = n - \sum_{i=1}^{f} w_i$
$\overrightarrow{w}^I$	Initial workpiece allocation
w_i	Number of workpieces at station i

is created, starting the described procedure from the beginning. This is repeated until all states are generated. In the following, each sub-procedure is described in detail.

5.4.1 Generation of Workpiece Allocations

The procedure WORKPIECE ALLOCATION generates all possible workpiece allocations $\overrightarrow{w}$ that can occur in a given configuration. This means, the algorithm creates all possible number series with M positions of values in the range of $w_i = 0, \ldots, d_i$ at each position i, and with the sum of all entries being equal to n. The value d_i indicates the maximum number of workpieces at station i, $d_i = \min(n, g_i)$. It is either restricted by the station capacity, denoted by g_i, with $g_i = b_i + s_i$, or by the number of workpieces in the system, n. The notation of this section is displayed in Table 5.6.

The algorithm starts with an initial workpiece allocation $\overrightarrow{w}^I$ and generates new workpiece allocations based on the current one. The principle of the algorithm is to allocate the workpieces as much as possible at the end of the production line without repeating earlier distributions. In the initial workpiece allocation $\overrightarrow{w}^I$, the workpieces are assigned to the stations with the highest possible station index subject to the station capacities. For example, in a five-station system with $n = 8$ workpieces and station capacities of $\overrightarrow{g} = (5, 5, 5, 5, 5)$, the initial workpiece allocation equals $\overrightarrow{w}^I = (0, 0, 0, 3, 5)^T$. Based on $\overrightarrow{w}^I$, one workpiece is moved one station further upstream and the rest is redistributed to the end of line.

The procedure consists of two parts, WORKPIECE ALLOCATION and its sub-procedure ARW (short for "allocation of remaining workpieces"). In WORKPIECE ALLOCATION, the partial workpiece allocation $w_1, \ldots, w_f$ is created. Based on this, ARW returns the second part of the workpiece allocation, $w_{f+1}, \ldots, w_M$. The index f indicates up to which station the workpiece allocation is fixed.

5.4.1.1 Procedure WORKPIECE ALLOCATION

The pseudo code of WORKPIECE ALLOCATION is given in Algorithm 8. In the procedure WORKPIECE ALLOCATION, the variable i^{UB} serves as a control parameter in the algorithm. It indicates the lowest station index containing a workpiece in the current allocation, see Eq. (5.21).

$$i^{UB} = \arg \min_i \{i = 1, \ldots, M \,|\, w_i > 0\}. \qquad (5.21)$$

i^{UB} serves as a control variable in WORKPIECE ALLOCATION. In the initial workpiece allocation of the example, i^{UB} equals 4.

WORKPIECE ALLOCATION tries to add a workpiece to the current workpiece allocation, where the last station is neglected. Hence, the addition-on-trial is started at station $k = M - 1$. If no further workpiece can be added to the considered station, k is decreased by 1 and the addition of a workpiece is tried again at that station. This is repeated until the addition is successful or until another decrease of k would mean that $k < i^{UB}$. The addition of a workpiece to station k depends on the fulfillment of two conditions:

1. $w_k + 1 \leq d_k$: An additional workpiece at station k must not violate the capacity restriction at station k.

2. $\left(\sum_{j=1}^{k} w_j \right) + 1 \leq n$: By adding a workpiece to station k, the sum of workpieces over stations $1, \ldots, k$ must not exceed the workpiece level n.

If both conditions hold true, a workpiece is added to station k, $w_k := w_k + 1$ and the workpiece distribution is fixed up to that station, i.e. the index f is set to k, $f := k$. Thus, the first part of the new workpiece allocation is composed. In the example, the initial value of k equals $k = M - 1 = 4$. In this example, the conditions for the addition of a workpiece at station k are fulfilled:

1. $w_4 + 1 = 4 \leq d_{M-1} = 5 \checkmark$
2. $\sum_{j=1}^{4} w_j + 1 = 4 \leq n = 8 \checkmark$

Since both conditions are met, the number of workpieces at station $k = 4$ is increased by 1 and the partial allocation $w_1, \ldots, w_f = (0, 0, 0, 4)$ is found. This partial workpiece distribution is passed on to ARW.

If the station capacity is reached ($w_k = d_k$) at the current upper bound of stations ($k = i^{UB}$), as is the case for $w_1, \ldots, w_k = (0, 0, 0, 5)$ with $i^{UB} = 4$, the station upstream of i^{UB} is opened up for the allocation of workpieces. This means that the upper bound of considered stations i^{UB} is decreased by 1 ($i^{UB} := i^{UB} - 1$). The highest station index of fixed workpieces is that newly opened station, which means f is set to the new upper bound, $f := i^{UB}$, and a workpiece is added to that station, $w_f := 1$. Next, the procedure ARW is called to allocate all other workpieces as furthest downstream as possible. The partial workpiece allocation in the example

---

Algorithm 8 Iterative generation of workpiece allocations

1: **procedure** WORKPIECE ALLOCATION$\{ M, n, \vec{d}, \vec{ph}^{NR}\}$
2:     **Initialization:** $\vec{w}^I, i^{UB}$
3:     **repeat**
4:         **for** $k = M - 1$ **to** i^{UB} **do**
5:             **if** $w_k + 1 \le d_k$ **and** $(\sum\limits_{j=1}^{k} w_j) + 1 \le n$ **then**
6:                 $w_k := w_k + 1$
7:                 $f := k$
8:                 **exit for**
9:             **else if** $k = i^{UB}$ **and** $w_k = d_k$ **then**
10:                $i^{UB} := i^{UB} - 1$
11:                **if** $i^{UB} = 0$ **then exit procedure**
12:                $f := i^{UB}$
13:                $w_f := 1$
14:            **end if**
15:        **end for**
16:        **Call procedure** ARW $\{M, n, \vec{g}, \vec{ph}^{NR}, f, \vec{w}\}$
17:        **Return** current $\vec{w}$
18:    **until** all possible $\vec{w}$ are generated
19: **end procedure**

---

equals at this step $w_1, \ldots, w_f = (0, 0, 1)$. In the example, the output of ARW equals $\vec{w} = (0, 0, 1, 2, 5)^T$.

The procedure WORKPIECE ALLOCATION is carried out until the station capacity of station $i = 1$ is reached and all workpieces are allocated furthest upstream. When the algorithm decreases i^{UB} to 0, all combinations have been created and the algorithm terminates.

5.4.1.2   Procedure ARW

The procedure ARW constitutes the second part of the generation of workpiece allocations. The pseudo code is presented in Algorithm 9. ARW allocates the remaining workpieces at the stations $f + 1, \ldots, M$ as furthest downstream as possible. The number of remaining workpieces n^R are obtained from $n^R = n - \sum\limits_{i=1}^{f} w_i$. The workpieces of the stations $f + 1, \ldots, M$ are initialized by zero. Starting with the last station ($k = M$), the workpiece-level at station k is incremented until the capacity limit is reached. Then, the station index is decremented, $k := k - 1$, and the new station k is loaded with workpieces in the same manner. This is done until all remaining workpieces are allocated. As a result, the workpiece allocation $\vec{w}$ is generated.

Figure 5.18 presents the first workpiece allocations and the corresponding values of the control variables for the given example. The first column reveals the output of WORKPIECE ALLOCATION. This first part of the workpiece allocation is marked

---

Algorithm 9 Allocation of remaining workpieces

1: **procedure** ARW$\{M, n, \overrightarrow{d}, \overrightarrow{ph}^{NR}, f, \overrightarrow{w}\}$
2:    $w_i := 0 \quad \forall\, f < i < M$
3:    $n^R := n - \sum_{i=1}^{f} w_i$
4:    $k := M$
5:    **repeat**
6:       **if** $w_k < w_k^{UB}$ **then**
7:          $w_k := w_k + 1$
8:          $n^R := n^R - 1$
9:       **else**
10:          $k := k - 1$
11:       **end if**
12:    **until** $n^R = 0$
13:    $S := S \bigcup \{s(\overrightarrow{q}, \overrightarrow{ph}^I, \overrightarrow{bs}^I)\}$
14:    **Call procedure** PHASES$\{\overrightarrow{w}, \overrightarrow{ph}^{NR}\}$
15: **end procedure**

---

output of Workpiece Allocation	resulting $\overrightarrow{w}$ output of ARW	current values of i^{UB} and k	check of conditions		consequences
	0003⑤	$i^{UB} = 4,\ k = 4$	(1) $w_k + 1 \le d_k$ ✓		
			(2) $\left(\sum_{j=1}^{k} w_j\right) + 1 \le n$ ✓	$\to f := k$	
					$w_f := w_f + 1$
☐0004x	0004④	$i^{UB} = 4,\ k = 4$	(1) ✓		
			(2) ✓	$\to f := k$	
					$w_f := w_f + 1$
☐0005x	0005③	$i^{UB} = 4,\ k = 4$	(1) $w_k + 1 \not\le d_k$	$\to i^{UB} := i^{UB} - 1$	
				$f := i^{UB}$	
				$w_f := 1$	
☐001xx	001㉕	$i^{UB} = 3,\ k = 4$	(1) ✓		
			(2) ✓		
☐0013x	0013④				

Fig. 5.18 Course of the generation of workpiece allocations

by a rectangular box. In the second column, the complete workpiece allocation, as a result from ARW, is listed. The part generated by ARW is indicated by an oval box. The current values of i^{UB} and k are provided in the next column. The check of the conditions is depicted in the fourth column. In the last column, the resulting consequences for the variables and the next workpiece allocation are presented. These depend on whether both conditions are met or not.

From each workpiece distribution, a state is derived: The queue lengths $\overrightarrow{q}$ result from the workpiece allocation, see Eq. (5.22). Depending on the workpiece allocation $\overrightarrow{w}$, initial indices of the active phases $\overrightarrow{ph}^I$ are assigned, see Eq. (5.23). The initial blocking states $\overrightarrow{bs}^I$ are given in Eq. (5.24).

Table 5.7 Workpiece allocations of the configuration of Table 5.8 with $n = 8$ and $g = (5, 5, 5, 5, 5)^T$

i	1	2	3	4	5	i	1	2	3	4	5
(1)	0	0	0	3	5						
(2)	0	0	0	4	4	(411)	5	1	0	0	2
(3)	0	0	0	5	3	(412)	5	1	0	1	1
(4)	0	0	1	2	5	(413)	5	1	0	2	0
(5)	0	0	1	3	4	(414)	5	1	1	0	1
(6)	0	0	1	4	3	(415)	5	1	1	1	0
(7)	0	0	1	5	2	(416)	5	1	2	0	0
(8)	0	0	2	1	5	(417)	5	2	0	0	1
(9)	0	0	2	2	4	(418)	5	2	0	1	0
(10)	0	0	2	3	3	(419)	5	2	1	0	0
...						(420)	5	3	0	0	0

Table 5.8 Five-station example of a CQN

i	1	2	3	4	5
μ_i	1	1	1	1	1
c_i^2	$0.\overline{3}$	0.5	$0.\overline{3}$	0.25	0.5
b_i	4	4	4	4	4
g_i	5	5	5	5	5
ph_i^{NR}	3	2	3	4	2

$$q_i = \begin{cases} w_i - 1 & \text{if } w_i > 0 \\ 0 & \text{if } w_i = 0 \end{cases} \quad \forall i \tag{5.22}$$

$$ph_i^I = \begin{cases} 1 & \text{if } w_i > 0 \\ 0 & \text{if } w_i = 0 \end{cases} \quad \forall i \tag{5.23}$$

$$bs_i^I = 0 \quad \forall i \tag{5.24}$$

Table 5.7 presents a configuration of a CQN with $n = 8$ customers. Table 5.8 shows several workpiece allocations according to the example of Table 5.7. The number of workpiece allocations amounts to 420. The initial workpiece allocation $\overrightarrow{w} = (0, 0, 0, 3, 5)^T$ composes the first state $s = (\overrightarrow{q}, \overrightarrow{ph}, \overrightarrow{bs})$ with the queue length $q_i = (0, 0, 0, 2, 4)^T$, the initial indices of the active phases $ph_i = (0, 0, 0, 1, 1)^T$, and the initial blocking statuses $bs_i = (0, 0, 0, 0, 0)^T$.

5.4.2　Generation of Phases

The procedure PHASES creates all possible combinations of active phases $\overrightarrow{ph}$ for a given workpiece allocation $\overrightarrow{w}$. The values of $\overrightarrow{ph}$ indicate the current active phase of the phase-type distribution at each station. A new combination of active phases is built upon the former. The indices of the active phases of the current iteration are denoted by ph_i^C.

Table 5.9 Notation

ph_i^C	Value of phases at station i in the current realization with $ph_i^C = 1, \ldots, ph_i^{NR}$ if $w_i \geq 1$ and $ph_i^C = 0$ if $w_i = 0$
ph_i^{LB}	Lower bound of the active-phase index at station i
$\overrightarrow{ph}^{LB}$	Vector of lower bounds of the active-phase indices at all stations
ph_i^{NR}	Number of phases in the phase-type distribution at station i
ph_i^{UB}	Upper bound of the active-phase index at station i
$\overrightarrow{ph}^{UB}$	Vector of upper bounds of the active-phase indices at all stations
i^{UB}	Lowest station index containing a workpiece
	$i^{UB} = \arg \min\limits_i \{i = 1, \ldots, M \,\|\, w_i > 0\}$

The variable ph_i^{UB} represents a control parameter in this procedure and denotes the maximal value of the active phase at station i. If there is no workpiece at station i, the upper bound of phases equals 0, i.e. no production is possible. Otherwise, if at least one workpiece is located at station i, the upper bound is determined by the number of phases ph_i^{NR} of the associated phase-type distribution, see Eq. (5.25).

$$ph_i^{UB} = \begin{cases} ph_i^{NR} & \text{if } w_i > 0 \\ 0 & \text{if } w_i = 0 \end{cases} \qquad \forall i \qquad (5.25)$$

The lower bound for the indices of the active phases is denoted by ph_i^{LB}. If station i does not contain any workpieces, it amounts to 0. However, if there are workpieces at station i, the first workpiece resides on the server and, therefore, the process is situated at least in the first phase:

$$ph_i^{LB} = \begin{cases} 1 & \text{if } w_i > 0 \\ 0 & \text{if } w_i = 0 \end{cases} \qquad \forall i \qquad (5.26)$$

As a result, the requirement of the algorithm is to generate all possible numerical series with M positions and all value combinations in the range of $ph_i^C = 1, \ldots,$ ph_i^{NR} if there is at least one workpiece at station i ($w_i > 0$), and $ph_i^C = 0$ otherwise. The notation is summarized in Table 5.9.

The procedure PHASES is given in Algorithm 10. The necessary input constitutes the lowest station index containing a workpiece, i^{UB}, see Eq. (5.21), the current queue lengths, given in Eq. (5.22), the initial indices of the active phases, provided in Eq. (5.23), and the initial blocking states, presented in Eq. (5.24).

The procedure PHASES iterates over the station index k from the last station $k = M$ down to i^{UB}. If station k contains at least one workpiece, i.e. $ph_k^{UB} > 0$, and if further the upper bound ph_k^{UB} is not reached yet, the current phase value at station k, ph_k^C, is increased by 1. After the increment, the iteration is exited. The current indices of the active phases ph_k^C for $k = 1, \ldots, M$ form the next combination of active phases. From that, another state (together with the given queue lengths and initial blocking statuses) is added to the state set. This procedure

---

Algorithm 10 Generation of phases for a given workpiece allocation

---

1: **procedure** PHASES$\{i^{UB}, \overrightarrow{q}, \overrightarrow{ph^I}, \overrightarrow{bs^I}\}$
2:   **repeat**
3:     **for** $k = M$ **to** i^{UB} **do**
4:       **if** $ph_k^{UB} > 0$ and $ph_k^C + 1 \le ph_k^{UB}$ **then**
5:         $ph_k^C += 1$
6:         **exit for**
7:       **else if** $ph_k^C = ph_k^{UB}$ **then**
8:         $ph_k^C = 1$
9:       **end if**
10:     **end for**
11:     $S += s(\overrightarrow{q}, \overrightarrow{ph}, \overrightarrow{bs^I})$
12:     Check if blocking at stations $i = 1, \ldots, M$ is possible
13:     **if** blocking is possible at any station **then**
14:       Generate vector of blockable stations
15:       **Call procedure** BLOCKING
16:     **end if**
17:   **until** $\overrightarrow{ph}^C = \overrightarrow{ph}^{UB}$
18: **end procedure**

Table 5.10 Parameters of the five-station example

i	1	2	3	4	5
$\mu(i)$	1	1	1	1	1
$c^2(i)$	$0.\overline{3}$	0.5	$0.\overline{3}$	0.25	0.5
$b(i)$	4	4	4	4	4
$ph^{NR}(i)$	3	2	3	4	2
$w(i)$	5	2	1	0	0
$q^C(i)$	4	1	0	0	0
$ph^I(i)$	1	1	1	0	0
$bs^I(i)$	0	0	0	0	0

is repeated until the upper bound of phases at station k is reached, i.e. $ph_k^C = ph_k^{UB}$. Then, the phase value is set back to 1 ($ph_k^C := 1$) and the phase value of the next station is increased by 1. This is done until all stations have been considered, which is true when $\overrightarrow{ph}^C = \overrightarrow{ph}^{UB}$. This way, all combinations are created.

We continue the example of Table 5.8 with the workpiece allocation $\overrightarrow{w} = (5, 2, 1, 0, 0)^T$. The resulting queue lengths, the initial phase statuses and the initial blocking states are displayed in Table 5.10. The resulting indices of the active phases are presented in Table 5.11. Each combination of active phases represents one state together with the current queue lengths q_i^C and initial blocking statuses bs_i^I. The number of different combinations of active phases amounts to 18 in this example.

For each combination of active phases $\overrightarrow{ph}^C$, the possibility of blocking is checked. If blocking is possible, a vector of the blockable stations, denoted by $\overrightarrow{bp}$, is generated and the procedure BLOCKING is called.

Table 5.11 Indices of the active phases of the example with input of Table 5.10

i	1	2	3	4	5		1	2	3	4	5		1	2	3	4	5
(1)	1	1	1	0	0	(7)	2	1	1	0	0	(13)	3	1	1	0	0
(2)	1	1	2	0	0	(8)	2	1	2	0	0	(14)	3	1	2	0	0
(3)	1	1	3	0	0	(9)	2	1	3	0	0	(15)	3	1	3	0	0
(4)	1	2	1	0	0	(10)	2	2	1	0	0	(16)	3	2	1	0	0
(5)	1	2	2	0	0	(11)	2	2	2	0	0	(17)	3	2	2	0	0
(6)	1	2	3	0	0	(12)	2	2	3	0	0	(18)	3	2	3	0	0

Table 5.12 Notation

bs_i^C	Value of blocking statuses at station i in the current realization, with $bs_i = 0$ if station i is not blocked and $bs_i = 1$ if station i is blocked
$\overrightarrow{bp}$	Vector indicating if a station can potentially be blocked, with $bp_i = 0$ if station i cannot be blocked and $bp_i = 1$ if station i can be blocked

5.4.3 Generation of Blocking Statuses

The procedure BLOCKING generates all combinations of blocking statuses over all stations for a given workpiece allocation and a given combination of active phases. It creates all possible number series with M positions of the values 0 and 1, if blocking is possible at station i, and of the value 0, if blocking is not possible at station i. Whether a blocking possibility exists depends on the queue lengths and the indices of the active phases. Station i becomes blocked with the rate of its current phase μ_{i,ph_i^C} if

1. $q_{i+1} = b_{i+1}$: the buffer at the subsequent station is full and
2. $ph_i^C = ph_i^{UB}$ (for all distributions) and additionally $ph_i^C = 1$ (if the Cox-2 distribution is present): the current phase represents the last processing phase.

If these conditions hold true, the Markov chain contains two states that are identical except for the blocking status of station i. There is one state in which the phase at station i is active (i.e. station i is not blocked) and another state in which station i is "blocked" while all other parameters stay the same. If blocking is possible at several stations, all combinations of blocking possibilities have to be considered.

The vector $\overrightarrow{bp}$ indicates whether blocking is possible at station i (Table 5.12). If the above conditions are met, the value for station i equals $bp_i = 1$, and $bp_i = 0$ otherwise. $\overrightarrow{bp}$ serves as a control parameter in the algorithm. The procedure BLOCKING equals the procedure PHASES except that $\overrightarrow{bp}$ constitutes the upper bound for the values instead of ph^{UB}. The pseudo code is presented in Algorithm 11.

To continue the example from above, consider the queue lengths $\overrightarrow{q} = (4, 4, 4, 0, 1)^T$ and the indices of the active phases $\overrightarrow{ph} = (3, 2, 1, 0, 2)^T$. These and all other data are given in Table 5.13. The conditions for blocking are fulfilled for station 1, 2 and 5: The last phase is active, $ph_1^{UB} = ph_1 = 3$, $ph_2^{UB} = ph_2 = 2$, $ph_5^{UB} = ph_5 = 2$ and the succeeding buffer is full, $q_2 = b_2 = 4$, $q_3 = b_3 = 4$, $q_1 = b_1 = 4$.

Algorithm 11 Generation of blocking for given queue lengths and given indices of the active phases

---

1: **procedure** BLOCKING$\{\overrightarrow{q}, \overrightarrow{ph}^{UB}, \overrightarrow{ph}^C, \overrightarrow{bs}^I, \overrightarrow{bp}, i^{UB}\}$
2:     **repeat**
3:         **for** $k = M$ **to** i^{UB} **do**
4:             **if** $bp_i = 1$ **then**
5:                 **if** $bs_k^C = 0$ **then**
6:                     $bs_k^C = 1$
7:                     **exit for**
8:                 **else if** $bs_k^C = 1$ **then**
9:                     $bs_k^C = 0$
10:                 **end if**
11:             **end if**
12:         **end for**
13:         $\mathcal{S} += s(\overrightarrow{q}, \overrightarrow{ph}, \overrightarrow{bs}^C)$
14:     **until** $\overrightarrow{bs}^C = \overrightarrow{bp}$
15: **end procedure**

Table 5.13 Values of the five-station example

i	1	2	3	4	5
$\mu(i)$	1	1	1	1	1
$c^2(i)$	$0.\overline{3}$	0.5	$0.\overline{3}$	0.25	0.5
$b(i)$	4	4	4	4	4
$ph(i)$	3	2	3	4	2
$w(i)$	5	5	5	0	2
$q(i)$	4	4	4	0	1
$ph^C(i)$	3	2	1	0	2
$bs^I(i)$	0	0	0	0	0
$bp(i)$	1	1	0	0	1

Table 5.14 Possible blocking statuses bs^C for the blocking possibilities $bp = (1, 1, 0, 0, 1)$

i	1	2	3	4	5
(1)	0	0	0	0	1
(2)	0	1	0	0	0
(3)	1	0	0	0	0
(4)	0	1	0	0	1
(5)	1	0	0	0	1
(6)	1	1	0	0	0
(7)	1	1	0	0	1

The vector of blocking possibilities equals $\overrightarrow{bp} = (1, 1, 0, 0, 1)^T$. Table 5.14 shows all possible blocking statuses bs^C for $\overrightarrow{bp}$. For the given queue lengths $\overrightarrow{q} = (4, 4, 4, 0, 1)^T$ and indices of the active phases $\overrightarrow{ph} = (3, 2, 1, 0, 2)^T$, seven states are conducted, see Table 5.14. The notation is provided in Table 5.16.

5.5 Transition Rates

The transition rate matrix of the Markov chain is determined after the state space is completely specified. This section addresses the transition rates both theoretically (in Sect. 5.5.1) and applied to the Markov-chain model of closed queueing networks (in Sect. 5.5.2). The transition rate matrix conforms with the global balance equations; they only differ by a rearrangement.

5.5.1 Derivation of the Global Balance Equations

In this section, the global balance equations are derived. This is the theoretical foundation of the continuous-time Markov chains (CTMC). We present the established formulas from theorems and properties through to the global balance equations. Section 5.5.1.1 presents the assumptions and requirements underlying the Markov-chain model of closed queueing networks. Using the stated properties and definitions as a pre-requisite, in Sect. 5.5.1.2, the proof of the global balance equations is presented. Examples are provided in Sect. 5.5.1.3.

The global balance equations specify the Markov chain by means of a linear set of equations. These equations are subject to the given transition rates and the unknown steady-state probabilities. By solving the system of linear equations, the steady-state probabilities of the continuous-time Markov chain are obtained.

Figure 5.19 provides an overview of the derivation of the global balance equations. The Markov property, the assumption of time-homogeneity, the theorem of total probability, and the independence assumption of random variables represent the basis of the derivation. From these statements, the Chapman Kolmogorov Equation can be deduced. The Markov property is used to state the relation between the transition probabilities p and the transition rates q. The Kolmogorov forward equation and the Kolmogorov backward equation are conducted using the relation between p and q, the Chapman Kolmogorov Equation, and the probability of the minimum of exponentially distributed random variables. The global balance equations are then derived from the Kolmogorov forward equation, together with the assumption of stationarity and the property of the exponential distribution about the minimum of its random variables.

5.5.1.1 Properties and Definitions

In the course of this section, firstly, the properties regarding the exponential distribution, and thereupon, the definitions concerning continuous-time Markov chains are presented. The presentation of each property is followed by its proof. The notation of this section is given in Table 5.15. The Markov property, also called lack-of-memory property, is stated in the following.

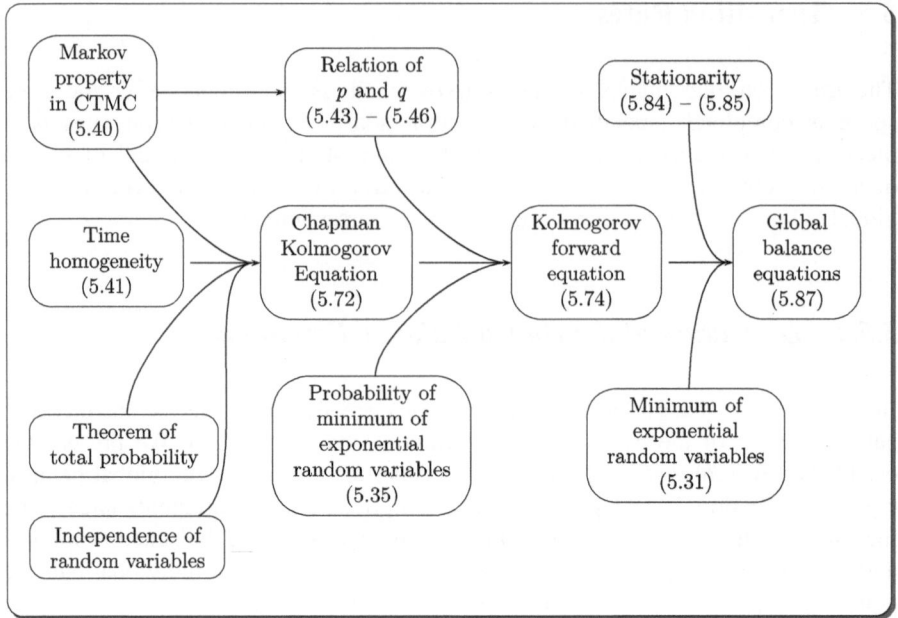

Fig. 5.19 Flowchart of the derivation of the global balance equations

Property 5.1 (Markov property). Let T denote an independent and identically distributed random variable following an exponential distribution with parameter μ. Then the Markov property, given in Eq. (5.27), holds.

$$P(T > s + t | T > s) = P(T > t) \qquad (5.27)$$

The Markov property states: Given that s time units have already passed since the last event, the time until the next event, t, has the same distribution as the entire time between two events has. This implies that the distribution of the remaining time t is independent of the elapsed time s. Applied to the service time, this means that the remaining service time follows the same exponential distribution as the service time for the complete process.[15]   □

Derivation of the Markov property: The density function of the exponentially distributed random variable T, $f(t)$, is provided in Eq. (5.28).[16]

$$f(t) = \mu \cdot e^{-\mu t} \qquad \text{for } t \geq 0 \qquad (5.28)$$

The Markov property is derived using that density function, see Eq. (5.29).

---

[15]See Ross (1996, p. 232).
[16]See Bolch et al. (2006, p. 20).

Table 5.15 Notation

$o(h)$	Any function converging faster to zero than its argument, $\lim\limits_{h\to 0}\frac{o(h)}{h}=0$
p_{ij}	Probability of transiting from state i to state j
$p_{ij}(0,h)$	Probability of transiting from state i to state j in the time interval $(0,h]$
$p_{ij}(h)$	Probability of transiting from state i to state j during time period h
$p'_{ij}(t)$	Marginal probability of transiting from state i to state j during time period t
π_i	Steady-state probability of being in state i
q_{ij}	Rate of transiting from state i to state j
q_i	Rate of leaving state i
T	Random variable
$X(t)$	Random variable of a stochastic process at time t

Table 5.16 Notation

bs_{si}	Value of blocking statuses at station i in state s
μ^τ_{sti}	Transition rate of the transition from state s to state t induced by station i
μ_{i,ph_i}	Exponential rate of the ph_i-th phase at station i
nrt_i	Index of exit after the first phase of station i
NRT_i	Number of exits after the first phase of station i
ph_{si}	Index of the active phases at station i in state s
q_{si}	Queue length at station i in state s
Q	Transition rate matrix
$Q(s,t)$	Transition rate from state s to state t
s	Origin state
t	Target state

$$P(T>s+t\mid T>s)=\frac{P(T>s+t,T>s)}{P(T>s)}=\frac{P(T>s+t)}{P(T>s)}$$

$$=\frac{e^{-\mu(s+t)}}{e^{-\mu s}}=e^{-\mu t}=P(T>t) \tag{5.29}$$

The left-hand side of the equation states the probability that the random variable T is greater than $s+t$ time units (or: that T exceeds t additional time units) given that s time units have already passed. This expression is transferred into the second term using the definition of the conditional probability. For the transformation into the third term the following applies: As the event $T>s+t$ involves the event $T>s$, the probability $P(T>s+t,T>s)$ can be reduced to $P(T>s)$. Applying the probability density function of the exponential distribution, the fourth term is set up. This equals the probability that T exceeds t time units, independent of the time passed. □

The following property, Property 5.2, states that the distribution of the minimum of several exponentially distributed random variables is also exponential. The simplicity of the result is of high value for the Markov-chain analysis.

Property 5.2. Let $T_1,\ldots,T_n$ represent n independent random variables and let each T_i for $i=1,\ldots,n$ be exponentially distributed with rate μ_i, $T_i\sim EXP(\mu_i)$.

Then, the distribution of the minimum of all random variables, $\min\{T_1, T_2, \ldots, T_n\}$, is exponentially distributed with the sum of all exponential rates:

$$\min\{T_1, T_2, \ldots, T_n\} \sim EXP(\mu_1 + \mu_2 + \ldots + \mu_n). \qquad \Box \qquad (5.30)$$

Derivation of Property 5.2.

$$
\begin{aligned}
P(\min\{T_1, T_2, \ldots, T_n\} > t) &= P(T_1 > t, T_2 > t, \ldots, T_n > t) \\
&= P(T_1 > t) \cdot P(T_2 > t) \cdot \ldots \cdot P(T_n > t) \\
&= \prod_{i=1}^{n} e^{-\mu_i t} \\
&= e^{-(\mu_1 + \mu_2 + \ldots + \mu_n) \cdot t} \\
&= P([T_1 + T_2 + \ldots + T_n] > t) \qquad (5.31)
\end{aligned}
$$

The step from lines 1 to 2 is possible due to the independence assumption of the random variables $T_1, \ldots, T_n$. The other steps are obtained by insertion and rearrangement. [17] $\Box$

In continuous-time Markov chains, the time spent in a state must be exponentially distributed. Properties 5.1 and 5.2 can be applied to the Markov-chain model of CQN; thus, it is shown that this requirement is fulfilled, see Property 5.3.

Property 5.3. The recurrence time of each state in a Markov chain that models a CQN in its phase-type representation is exponentially distributed. $\Box$

Derivation of Property 5.3. Let $T_{i,v}$ denote the time of the active phase at station v in state i. Further, let $T_{i,u}^{r}$ indicate the remaining time of the active phase at station u in the same state i. At the time-instant of entering state i, processing has either already started or processing starts at the time-instant of entering that state. The start of a new phase occurs only at one station. As a side note, it is the station which induced the transition into the considered state i.

The system remains in a state until the first of all active phases is finished. Hence, the recurrence time of state i, denoted by T_i, equals the minimum of all residual service times of the current active phases at stations $u = 1, \ldots, M$, with $u \neq v$, and the complete service time of the active phase at station v. Station v denotes the station which has started the current active phase at the time-instant of entering state i. Therefore, the time spent in state i, T_i, equals

$$T_i = \min_{u=1,\ldots,M, u \neq v} \{T_{i,u}^{r}, T_{i,v}\}. \qquad (5.32)$$

---

[17]See Ross (1997, p. 243).

Each phase is exponentially distributed. Therefore, the residual time spent in a phase is distributed as the complete time in phase, $T_{i,u}^r \sim T_{i,u}$, see Eq. (5.27). Furthermore, the minimum of exponentially distributed random variables is exponentially distributed as well, namely by the sum of the exponential rates, see Eq. (5.31). Thus, it holds that

$$T_i \sim EXP\left(\sum_{u=1}^M q_{i,u}\right) \forall i. \qquad \Box \qquad (5.33)$$

where $q_{i,u}$ denotes the rate of the active phase at station u in state i with $q_{i,u} = 0$ if the station is inactive. In summary, the recurrence time of each state is exponentially distributed first because the residual service times are exponentially distributed and second because the minimum of the residual service times is exponentially distributed as well.

Applying Properties 5.1 and 5.2 to the three-station example of Fig. 5.9,[18] the following holds true: The recurrence time of state (a) is distributed as $\min(T_{1,1}, T_{2,2}, T_{3,1}) \sim EXP(\mu_{1,1} + \mu_{2,2} + \mu_{3,1})$. The expected residence time of state (a) equals $E[T_{(a)}] = 1/(\mu_{1,1} + \mu_{2,2} + \mu_{3,1})$.

Next to the time spent in a state, the probability for a particular transition from one state to another is of interest. Assume e.g. the residence in state (a) in the setting of Fig. 5.9. In order to reach state (b) from state (a), phase 1 at station 3 must be finished before any other active phase. According to Property 5.4, the probability for this event equals $P(\min\{T_{1,1}, T_{2,2}, T_{3,1} = T_{3,1}\}) = \mu_{3,1}/(\mu_{1,1} + \mu_{2,2} + \mu_{3,1})$.

Property 5.4. Let $T_1, T_2, \ldots, T_n$ represent n independent random variables, which are exponentially distributed with rate $\mu_1, \mu_2, \ldots, \mu_n$. Then, the probability that the realization of T_i corresponds to the smallest realization of all random variables equals[19]:

$$P(T_i = \min\{T_1, T_2, \ldots, T_n\}) = \frac{\mu_i}{\mu_1 + \mu_2 + \ldots + \mu_n}. \qquad \Box \qquad (5.34)$$

Derivation of Property 5.4. [20]

$$P(T_i = \min\{T_1, T_2, \ldots, T_n\}) = P(T_i < T_j \; \forall j \neq i)$$

$$= \int_0^\infty P(T_i < T_j | T_i \leq t, \; \forall j \neq i) \cdot P(T_i \leq t) \, dt$$

---

[18] See page 72.
[19] See Ross (1997, p. 243).
[20] See Ross (1997, p. 243).

$$= \int_0^\infty P(t < T_j, \forall j \neq i) \cdot \mu_i e^{-\mu_i t} \, dt$$

$$= \int_0^\infty \mu_i e^{-\mu_i t} \prod_{j \neq i}^n P(T_j > t) \, dt$$

$$= \int_0^\infty \mu_i e^{-\mu_i t} \prod_{j \neq i}^n e^{-\mu_j t} \, dt$$

$$= \mu_i \int_0^\infty e^{-(\mu_1 + \mu_2 + \ldots + \mu_n)t} \, dt$$

$$= \mu_i \left[\frac{-e^{-(\mu_1 + \mu_2 + \ldots + \mu_n)t}}{\mu_1 + \mu_2 + \ldots + \mu_n} \right]_0^\infty$$

$$= \frac{\mu_i}{\mu_1 + \mu_2 + \ldots + \mu_n} \qquad \square \qquad (5.35)$$

In the following, the rate of departing from state i is denoted by q_i. A particular transition from state i to state j is given by the transition rate q_{ij}. According to the distribution of the minimum of exponential rates, see Eq. (5.31), it holds that q_i equals the sum of all departure rates into any state j:

$$q_i = \sum_{j \in \mathcal{S}} q_{ij}. \qquad (5.36)$$

The probability of transiting from state i into a particular state j, denoted by p_{ij}, equals according to Eq. (5.34)

$$p_{ij} = \frac{q_{ij}}{q_i}. \qquad (5.37)$$

By rearrangement of (5.37), the transition rate q_{ij} can be expressed by

$$q_{ij} = p_{ij} \cdot q_i \qquad (5.38)$$

Subsequently, the definition of a stochastic process, of a continuous-time Markov chain, and of the concept of time-homogeneity are presented.

Definition 5.1. A **stochastic process** $\{X_t : t \in T\}$ constitutes a collection of random variables. The set of all possible values of X_t is known as the state space $\mathcal{S}$ of the stochastic process. Each random variable X_t is indexed by the parameter $t \in T$. t usually conforms with the time parameter if $T \subseteq \mathbb{R}_+ = [0, \infty)$. If T is a countable set, the stochastic process takes place in discrete time. If T is continuous, the stochastic process represents a continuous-time process.[21]   □

Here, we focus on continuous-time stochastic processes. For the next definition, the time-dependent transition probability is introduced:

$$p_{ij}(0,t) = P(X(t) = j | X(0) = i). \tag{5.39}$$

$p_{ij}(0,t)$ constitutes the probability of transiting from state i to state j during the time interval $[0,t)$. It can also be expressed by the probability of residing in state j at the time-instant t, given that the system resides in state i at the time-instant 0, denoted by $P(X(t) = j | X(0) = i)$.

Definition 5.2. A stochastic process X_t constitutes a **continuous-time Markov chain** if for arbitrary points in time $t_i \in \mathbb{R}_0^+$ with $i \in \mathbb{N}$ denoting the i-th transition and $0 = t_0 < t_1 < \ldots < t_n$ and for all states s_i of the state set $\mathcal{S}$ the following relation holds:

$$P(X_{t_{n+1}} = s_{n+1} | X_{t_n} = s_n, X_{t_{n-1}} = s_{n-1}, \ldots, X_{t_0} = s_0)$$
$$= P(X_{t_{n+1}} = s_{n+1} | X_{t_n} = s_n) \tag{5.40}$$

Equation (5.40) expresses the Markov property for continuous-time Markov chains. It states that the probability of entering state s_{n+1} next, at an arbitrary point in time t_{n+1}, given the present and past states, only depends on the present state s_n.[22]   □

Definition 5.2 implies that a continuous-time Markov chain is a stochastic process with the properties

- That the residence time of each state is exponentially distributed and
- That the probability of transiting from any state i to the state j equals $\sum_{i, i \neq j} p_{ij} = 1$.[23]

In the steady-state analysis of continuous-time Markov chains, we assume the transition probabilities to be independent of the points in time and only depend on the amount of time. This is expressed in Definition 5.3.

Definition 5.3. Let $p_{ij}(u,v)$ denote the probability of a transition from state i to state j during time-period $[u,v)$. If the transition probability $p_{ij}(u,v)$ is independent of the points in time u and v and only depends on the time period $t = v - u$, the Markov chain is said to possess **time-homogeneous** or **stationary transition**

---

[21]Bolch et al. (2006, p. 52) and Ross (1997, p. 77).

[22]See Bolch et al. (2006, p. 65).

[23]See Ross (1996, p. 232).

probabilities. The transition probability is then denoted by $p_{ij}(t)$. In case of time-homogeneity, Eq. (5.41) holds.[24]

$$p_{ij}(u,v) = p_{ij}(0,t) = P(X_t = j \,|\, X_0 = i) = p_{ij}(t)$$

$$\forall i,j \in S \text{ and } \forall u,v,t \in T \quad \square \quad (5.41)$$

The succeeding property, Property 5.5, describes the relation between the steady-state probabilities and the transition rates, the global balance equations. These, together with the normalizing equation in (5.90), constitute a set of linear equations that allows the computation of the steady-state probabilities.

Property 5.5. In continuous-time Markov chains with discrete state space and time-homogeneity, the following relation between the transition rates and the steady-state probabilities holds, see Eq. (5.42). These are the so-called **global balance equations**.

$$\sum_{\substack{i \in S \\ i \neq j}} q_{ij} \pi_i = \sum_{\substack{i \in S \\ i \neq j}} q_{ji} \pi_j \qquad \forall j \in S \qquad (5.42)$$

q_{ij} denotes the transition rate from state i to state j, and π_i indicates the steady-state probability of residing in state i. S labels the state space.[25] $\square$

5.5.1.2   Proof of the Global Balance Equations

In this section, the global balance equations are derived. The proof is divided into the four following parts:

1. Relation of the transition probabilities and the transition rates
2. Chapman-Kolmogorov Equation
3. Kolmogorov forward and backward equations
4. Global balance equations

1. Relation of the transition probabilities and the transition rates

The objective of this part is to express the transition rates from state i to state j, q_{ij} $\forall i,j \in S$, subject to the transition probabilities from state i to state j during time period h, $p_{ij}(h)$ $\forall i,j \in S$ and vice versa[26]:

$$p_{ij}(h) = q_{ij} \cdot h + o(h) \qquad \forall i,j \in S, i \neq j \qquad (5.43)$$

$$p_{ii}(h) = 1 - q_i \cdot h + o(h) \qquad \forall i \in S \qquad (5.44)$$

[24]See Bolch et al. (2006, p. 53).

[25]See Bolch et al. (2006, p. 69).

[26]See Ross (1997, p. 316).

$$q_{ij} = \lim_{h \to 0} \frac{p_{ij}(h)}{h} \qquad \forall i, j \in \mathcal{S}, i \neq j \qquad (5.45)$$

$$q_i = \lim_{h \to 0} \frac{1 - p_{ii}(h)}{h} \qquad \forall i \in \mathcal{S}. \qquad (5.46)$$

The function $o(h)$ represents any function that converges faster to zero than its argument h, which means $\lim_{h \to 0} \frac{o(h)}{h} = 0$.[27] It does not matter from which direction and how fast it converges. This expression is useful because it simplifies the terms to the parts that are needed, namely the limiting properties.

The starting point to derive these relations represents the probability of being in state i at time h under the condition of being in state j at time 0. It is denoted by $P(X(h) = j | X(0) = i)$ or, in short, by $p_{ij}(0, h)$. This event takes place if

- Exactly one transition occurs from state i to state j in the time-interval [0,h) or
- At least two transitions occur in the time-interval [0,h), whereby the system starts in state i at time 0, transits into any other state(s), and resides in state j at the time-instant h.

Since the residence times are exponentially distributed, see Eq. (5.33), the number of transitions constitutes a Poisson process.[28] Therefore, the probability of starting in state i at time $t = 0$ and transiting within h time units into state j can be expressed by the number of transitions. This probability is, according to the listing above, decomposed into the following parts[29]:

$$\begin{aligned} P(X(h) &= j | X(0) = i) \\ &= P(NT(0, h) = 1 | X(0) = i) \cdot p_{ij} \\ &\quad + P(NT(0, h) \geq 2 | X(0) = i) \\ &\qquad \cdot P(\text{transition from } i \text{ to } j \text{ with visits in arbitrary states in between}) \end{aligned}$$

$$(5.47)$$

$NT(0, h)$ denotes the number of transitions within the time-interval [0,h). In the following, we investigate each element of Eq. (5.47). We denote the time spent in state $i \in \mathcal{S}$, i.e. the recurrence time of state i, by T_i. It is exponentially distributed with parameter q_i: $T_i \sim EXP(q_i)$ with $0 < q_i < \infty$.

The distribution of the number of transitions is determined via the recurrence time in a state. The probability that the recurrence time of state i, T_i, exceeds h time-units equals

$$P(T_i > h) = 1 - F(h) = 1 - (1 - e^{-q_i \cdot h}) = e^{-q_i \cdot h}. \qquad (5.48)$$

---

[27] For $o(h)$ holds e.g. $o(h) + o(h) = o(h)$, $o(h) \cdot o(h) = o(h)$, and $c \cdot o(h) = o(h)$, see Ross (1997, p. 251f).

[28] See Ross (1996, pp. 251f).

[29] See Ross (1996, p. 492).

Expanding the exponential function of (5.48) in a Taylor series,[30] we obtain[31]

$$P(T_i > h) = e^{-q_i \cdot h}$$

$$= 1 - q_i \cdot h + \frac{(q_i \cdot h)^2}{2!} - \frac{(q_i \cdot h)^3}{3!} + \dots$$

$$= 1 - q_i \cdot h + o(h) \tag{5.49}$$

$$\Leftrightarrow P(T_i \le h) = q_i \cdot h + o(h) \tag{5.50}$$

Equation (5.49) expresses the probability that the residence time in state i is larger than h time units. This corresponds to the probability that no transition takes place in the time interval [0,h):

$$P(NT(0, h) = 0 | X(0) = i) = P(T_i > h) = 1 - q_i \cdot h + o(h). \tag{5.51}$$

Equation (5.50) indicates the probability that the residence time in state i maximally amounts to h. It can be used to express the probability that at least one transition occurs during [0,h):

$$P(NT(0, h) \ge 1 | X(0) = i) = P(T_i \le h) = q_i \cdot h + o(h). \tag{5.52}$$

The probability of two or more transitions within the interval $[0, h)$ can be stated as the probability that the residence time of two succeeding states i and j endures at most h time units.

$$P(NT(0, h) \ge 2 | X(0) = i) = P(T_i + T_j \le h). \tag{5.53}$$

Equation (5.53) can be transformed as follows[32]:

$$P(T_i + T_j \le h) \le P(T_i \le h) \cdot P(T_j \le h)$$

$$= (q_i \cdot h + o(h)) \cdot (q_j \cdot h + o(h))$$

$$= q_i \cdot q_j \cdot h^2 + o(h)$$

$$= o(h). \tag{5.54}$$

As a side note, since $\lim\limits_{h \to 0} \frac{q_i \cdot q_j \cdot h^2}{h} = 0$, it follows that $q_i \cdot q_j \cdot h^2 = o(h)$.

---

[30]The exponential function can be characterized by a Taylor series:

$e^x = \sum\limits_{n=0}^{\infty} \frac{x^n}{n!} = 1 + x + \frac{x^2}{2!} + \frac{x^3}{3!} + \dots$

[31]See Allen (1990, p. 203).

[32]See Ross (1996, p. 492).

Exactly one transition occurs in the interval $[0,h)$ if at time h, state i is already finished; however, the subsequent state j still endures. The probability of exactly one transition between 0 and h (given that the Markov chain resides in state i at time 0) is expressed using Eqs. (5.50) and (5.54).[33]

$$P(NT(0,h) = 1|X(0) = i)$$
$$= \sum_{j \neq i \in \mathcal{S}} P(T_i \leq h < T_i + T_j | X(0) = i)$$
$$= P(T_i + T_j > h | X(0) = i) - P(T_i \leq h | X(0) = i)$$
$$= P(T_i + T_j > h) - P(T_i \leq h)$$
$$= (1 - o(h)) - (1 - q_i \cdot h + o(h))$$
$$= q_i \cdot h + o(h) \tag{5.55}$$

The probability of two or more transitions by time h given that the system resides in state i at time 0 equals the probability that h exceeds or equals the recurrence time of that state i and of the successor state, which is any state $j \neq i$, see Eq. (5.56).[34]

$$P(NT(0,h) \geq 2|X(0) = i)$$
$$= \sum_{j \neq i} P(NT(0,h) \geq 2|X(0) = i, \text{ transition to } j) \cdot p_{ij}$$
$$= \sum_{j \neq i} P(T_i + T_j \leq h) \cdot p_{ij}$$
$$= \sum_{j \neq i} o(h) \cdot p_{ij}$$
$$= o(h) \tag{5.56}$$

p_{ij} constitutes the probability of transiting from state i to state j at all, regardless of the time h. p_{ij} is unknown. As it is independent of h, it can be treated as a constant, and as such it is absorbed by $o(h)$.

Employing the equations derived above on (5.47), one obtains the following expression[35]:

$$P(X(h) = j|X(0) = i)$$
$$= P(NT(0,h) = 1|X(0) = i) \cdot p_{ij}$$

[33] See Allen (1990, pp. 203f).
[34] See Ross (1997, p. 637).
[35] See Ross (1996, pp. 492f).

$$+ P(NT(0, h) \geq 2|X(0) = i)$$

$$\cdot P(\text{any number of transitions starting in } i \text{ and arriving in } j)$$

$$= [q_i \cdot h + o(h)] \cdot p_{ij} + o(h)$$

$$\cdot P(\text{any number of transitions starting in } i \text{ and arriving in } j)$$

$$= q_i \cdot h \cdot p_{ij} + o(h). \tag{5.57}$$

Equation (5.57) expresses the desired transition probability from state i to state j during [0,h). Thereby, Eq. (5.43) has been derived.

The event to be in state i at time h, given that the system resides in the same state i at time 0, is accomplished by either staying in state i during the complete time interval or by transiting twice or more often and having returned into state i by time h.[36] In accordance with the expressions obtained above, we state the following equation by applying (5.49) and (5.56)[37]:

$$P(X(h) = i|X(0) = i)$$

$$= P(NT(0, h) = 0|X(0) = i) + P(NT(0, h) \geq 2|X(0) = i)$$

$$\cdot P(\text{any number of transitions starting in } i \text{ and arriving in } i)$$

$$= P(T_i > h|X(0) = i) + P(T_i + T_j \leq h|X(0) = i)$$

$$= [1 - q_i \cdot h + o(h)] + o(h)$$

$$= 1 - q_i \cdot h + o(h) \tag{5.58}$$

Up to this point, the following expressions have been derived:

$$p_{ij}(0, h) = q_i \cdot h \cdot p_{ij} + o(h) \tag{5.59}$$

$$p_{ii}(0, h) = 1 - q_i \cdot h + o(h) \tag{5.60}$$

Using $q_{ij} = p_{ij} \cdot q_i$ from Eq. (5.38), Eq. (5.59) can be stated as

$$p_{ij}(0, h) = q_{ij} \cdot h + o(h). \tag{5.61}$$

Employing the assumption of time-homogeneity, which is specified in (5.41), on Eqs. (5.60) and (5.61), the relations between the transition probabilities and the transition rates independent of the starting point are obtained:

$$p_{ij}(h) = q_{ij} \cdot h + o(h) \tag{5.62}$$

$$p_{ii}(h) = 1 - q_i \cdot h + o(h). \tag{5.63}$$

---

[36]See Ross (1997, pp. 252ff).
[37]See Ross (1997, pp. 637ff).

Equations (5.62) and (5.63) express the transition probabilities within h time units. Since a transition may occur at any time-instant, we are interested in a time-interval that is as small as possible. Therefore, we let $h \to 0$. By rearrangement, we obtain the transition rates subject to the transition probabilities under the assumption of time-homogeneity[38]:

$$q_{ij} = \lim_{h \to 0} \frac{p_{ij}(h)}{h}, \qquad i \neq j \tag{5.64}$$

$$q_i = \lim_{h \to 0} \frac{1 - p_{ii}(h)}{h} \tag{5.65}$$

As a result of the expressions obtained above, a transition rate can be thought of as a transition probability per time unit, in analogy to the speed that expresses the distance per time unit.[39] The rate q_i constitutes the rate of departing from state i to any state $j \in S$, including state i, see Eq. (5.36). Hence, it is expressed by means of the counter-probability of staying in state i. The rate q_{ij} expresses the transition rate from state i to state j, with $j \neq i$. Since the Markov chain has to be in any state, either by transiting into any other state, i.e. $j \neq i$, or by staying in the current state, i.e. $j = i$, it holds that

$$\sum_{j \in S} p_{ij}(h) = 1 \qquad \forall i \in S, \forall h \in T. \tag{5.66}$$

From Eqs. (5.64) to (5.66) follows[40]

$$\sum_{j \in S} q_{ij} = 0 \qquad \forall i \in S. \tag{5.67}$$

2. Chapman-Kolmogorov Equation

The Chapman-Kolmogorov Equation for continuous-time Markov chains is given in Eq. (5.68). It states that the probability of transiting from state i to state j within a time-interval of $t + h$ time units equals the probability of transiting from state i to an intermediate stop into any other state k in t time-units and from there to state j within h time-units.[41]

$$p_{ij}(t + h) = \sum_{k \in S} p_{ik}(t) \cdot p_{kj}(h) \qquad 0 \leq t < t + h \tag{5.68}$$

---

[38] See Bolch et al. (2006, p. 66).
[39] See Stewart (2009, p. 254).
[40] See Bolch et al. (2006, p. 66).
[41] See Bolch et al. (2006, p. 67).

Proof of the Chapman-Kolomogorov Equation:

$$p_{ij}(t+h) = P(X(t+h) = j \,|\, X(0) = i) \tag{5.69}$$

$$= \sum_{k \in S} P(X(t+h) = j \,|\, X(t) = k, X(0) = i)$$

$$\cdot P(X(t) = k \,|\, X(0) = i) \tag{5.70}$$

$$= \sum_{k \in S} P(X(t+h) = j \,|\, X(t) = k) \cdot P(X(t) = k \,|\, X(0) = i) \tag{5.71}$$

$$= \sum_{k \in S} P(X(h) = j \,|\, X(0) = k) \cdot P(X(t) = k \,|\, X(0) = i) \tag{5.72}$$

$$= \sum_{k \in S} p_{ik}(t) \cdot p_{kj}(h) \tag{5.73}$$

From Eqs. (5.69) to (5.70), the theorem of total probability is applied. From Eqs. (5.70) to (5.71) the Markov property is used as shown in Eq. (5.40). The step from Eqs. (5.71) to (5.72) and from Eqs. (5.72) to (5.73) is due to time-homogeneity, see Eq. (5.41). The probabilities in Eq. (5.73) may be multiplied because the recurrence times of all states are independent random variables.[42]

□

3. Kolmogorov forward and backward equations

The Kolmogorov backward equation is given by

$$p_{ij}'(t) = \frac{\delta p_{ij}(t)}{\delta t} = \sum_{\substack{k \in S \\ k \neq i}} q_{ik} \cdot p_{kj}(t) - q_i \cdot p_{ij}(t) \qquad \forall i, j \in S \tag{5.74}$$

The Kolmogorov forward equation constitutes

$$p_{ij}'(t) = \frac{\delta p_{ij}(t)}{\delta t} = \sum_{\substack{k \in S \\ k \neq j}} q_{kj} \cdot p_{ik}(t) - q_j \cdot p_{ij}(t) \qquad \forall i, j \in S \;[43] \tag{5.75}$$

Proof of the Kolmogorov backward equation:

The Chapman-Kolomogorov Equation, as defined in Eq. (5.68) equals

$$p_{ij}(t+h) = \sum_{k \in S} p_{ik}(h) \cdot p_{kj}(t) \qquad 0 \leq t < t + h. \tag{5.76}$$

---

[42] See Ross (1997, p. 317).
[43] See Bolch et al. (2006, p. 67), Ross (1996, p. 242, 1997, p. 320).

The expressions derived in Eq. (5.62),

$$p_{ii}(h) = 1 - q_i \cdot h + o(h),\tag{5.77}$$

and Eq. (5.63),

$$p_{ik}(h) = q_i \cdot p_{ik} \cdot h + o(h),\tag{5.78}$$

are applied on the Chapman-Kolomogorov Equation such that

$$p_{ij}(t + h) = (1 - q_i \cdot h + o(h)) \cdot p_{ij}(t)$$
$$+ \sum_{k \neq i} (q_i \cdot p_{ik} \cdot h + o(h)) \cdot p_{kj}(t)\tag{5.79}$$

is obtained. Subtracting $p_{ij}(t)$ on both sides leads to

$$p_{ij}(t + h) - p_{ij}(t) = -q_i \cdot p_{ij}(t) \cdot h$$
$$+ \sum_{k \neq i} p_{kj}(t) \cdot q_i \cdot p_{ik} \cdot h + o(h).\tag{5.80}$$

Upon dividing Eq. (5.80) by h and using $q_i \cdot p_{ik} = q_{ik}$ (see Eq. 5.38) it follows that

$$\frac{p_{ij}(t + h) - p_{ij}(t)}{h} = \sum_{k \neq i} q_{ik} \cdot p_{kj}(t) - q_i \cdot p_{ij}(t) + \frac{o(h)}{h}.\tag{5.81}$$

Letting h approach zero, $h \to 0$, this yields

$$\lim_{h \to 0} \frac{p_{ij}(t + h) - p_{ij}(t)}{h} =: p'_{ij}(t)$$
$$p'_{ij}(t) = \sum_{k \neq i} q_{ik} \cdot p_{kj}(t) - q_i \cdot p_{ij}(t).\,^{44} \qquad \square \tag{5.82}$$

Proof of the Kolmogorov forward equation:

The proof of the Kolmogorov forward equation is carried out analogously to that of the backward equation. Starting with the Chapman-Kolmogorov Equation

$$p_{ij}(t + h) = \sum_{k \in S} p_{ik}(h) \cdot p_{kj}(t) \qquad 0 \leq t < t + h\tag{5.83}$$

---

[44] See Ross (1997, pp. 317f).

and applying again Eqs. (5.62) and (5.63) on the Chapman-Kolomogorov Equation, one obtains

$$p_{ij}(t+h) = p_{ij}(h) \cdot (1 - q_j \cdot h + o(h))$$

$$+ \sum_{k \neq j} p_{ik}(h) \cdot (q_k \cdot p_{kj} \cdot h + o(h)). \tag{5.84}$$

Using $q_k \cdot p_{kj} = q_{kj}$ and performing the same transformations as for the Kolmogorov backward equation, one receives the Kolmogorov forward equation.[45]

4. Global balance equations

In order to calculate the performance measures, the steady-state probabilities are needed. A Markov chain is situated in steady state if the state probabilities do not depend on a certain point in time t. In order to derive the state probabilities in stationarity, we let t approach infinity, $t \to \infty$:

$$\pi_j := \lim_{t \to \infty} p_{ij}(t) \qquad \forall j \in \mathcal{S}. \tag{5.85}$$

π_j denotes the unconditional probability to be in state j. Equation (5.85) expresses: No matter in which state the system starts at time 0, the probability of residing in state j at time t approaches π_j as t approaches infinity.

Stationarity further requires that the state probabilities do not change over time. Mathematically expressed, this is[46]

$$\lim_{t \to \infty} \frac{\partial \pi_j(t)}{\partial t} := 0 \qquad \forall j \in \mathcal{S} \tag{5.86}$$

To obtain the global balance equations for the π_j, the Kolmogorov forward equation, see (5.75), is applied. Further, by letting t approach infinity, according to Eq. (5.85), and by using the fact that $q_j = \sum_{i \in S} q_{ji}$, the following expression is obtained[47]:

$$\lim_{t \to \infty} p'_{ij}(t) = \lim_{t \to \infty} \left[\sum_{\substack{k \in S \\ k \neq i}} q_{kj} \cdot p_{ik}(t) - q_j \cdot p_{ij}(t) \right]$$

[45]See Ross (1997, pp. 319f).
[46]See Bolch et al. (2006, p. 69).
[47]See Ross (1997, p. 322).

$$= \sum_{\substack{k \in \mathcal{S} \\ k \neq i}} q_{kj} \cdot \pi_k - q_j \cdot \pi_j$$

$$= \sum_{\substack{k \in \mathcal{S} \\ k \neq i}} q_{kj} \cdot \pi_k - \sum_{\substack{i \in \mathcal{S} \\ i \neq j}} q_{ji} \cdot \pi_j. \tag{5.87}$$

By applying condition (5.86) on Eq. (5.87), the global balance equations are derived:

$$\sum_{\substack{k \in \mathcal{S} \\ k \neq j}} q_{kj} \pi_k - \sum_{\substack{i \in \mathcal{S} \\ i \neq j}} q_{ji} \pi_j = 0. \qquad \forall j \in \mathcal{S}. \tag{5.88}$$

Equation (5.88) refers to each state $j \in \mathcal{S}$. The first term represents the arrival to state j from any state $k \in \mathcal{S}$. The parameter q_{kj} states the transition rate from state k to state j. The second term constitutes the departure from state j into any state $i \in \mathcal{S}$. Analogously, q_{ji} indicates the transition rate from state j to state i.

The global balance equations can be reformulated in matrix form. Let Q denote the transition rate matrix containing all transition rates q_{ij} $\forall i, j \in \mathcal{S}$. The vector P entails the steady-state probabilities π_j $\forall j \in \mathcal{S}$. With this notation, the system of equations (5.88) is stated in matrix form by

$$Q \cdot P = 0. \tag{5.89}$$

The steady-state probabilities are obtained from the solution of the global balance equations together with the normalizing equation. That equation conditions the sum of all steady-state probabilities of the Markov chain to add up to 1. It is given in Eq. (5.90).[48]

$$\sum_{i \in \mathcal{S}} \pi_i = 1 \tag{5.90}$$

We stated that the recurrence time of each state is exponentially distributed because it equals the sum of all active exponential phases (which is exponentially distributed as well). These rates do not change over time. Therefore, the Markov chain can be modeled time-homogeneously. The discrete and finite state space results from the finite capacity and given number of customers in the network. As a result, closed queueing networks with phase-type distributed processing times and finite buffer space can be modeled as homogeneous Markov chains in continuous time with discrete and finite state space. Therefore, the global balance equations can be applied.

---

[48]See Bolch et al. (2006, p. 69).

5.5.1.3   Examples

In the following, Examples 1,2, and 3 are continued by stating the global balance equations that correspond to the graphical representations of the Markov chains. The steady-state probability of state s is denoted by $\pi(s)$.

According to the global balance equations, see (5.88), the Markov chain of Fig. 5.15 (Example 1) leads to the set of equations given below.

$$(\mu_1 + \mu_2) \cdot \pi(2,2,0) = \mu_1 \cdot \pi(2,1,B_3(1)) + \mu_3 \cdot \pi(1,2,1)$$

$$(\mu_1 + \mu_2 + \mu_3) \cdot \pi(2,1,1) = \mu_1 \cdot \pi(2,0,B_3(2)) + \mu_2 \cdot \pi(2,2,0)$$
$$+ \mu_3 \cdot \pi(1,1,2)$$

$$(\mu_1 + \mu_2 + \mu_3) \cdot \pi(1,2,1) = \mu_1 \cdot \pi(2,1,1) + \mu_2 \cdot \pi(B_1(2),2,0)$$
$$+ \mu_3 \cdot \pi(0,2,2)$$

$$(\mu_1 + \mu_2 + \mu_3) \cdot \pi(1,1,2) = \mu_1 \cdot \pi(2,0,2) + \mu_2 \cdot \pi(1,2,1)$$
$$+ \mu_3 \cdot \pi(0,B_2(2),2)$$

$$(\mu_2 + \mu_3) \cdot \pi(0,2,2) = \mu_1 \cdot \pi(1,1,2) + \mu_2 \cdot \pi(B_1(1),2,1)$$

$$(\mu_1 + \mu_3) \cdot \pi(2,0,2) = \mu_2 \cdot \pi(2,1,1) + \mu_3 \cdot \pi(1,B_2(1),2)$$

$$(\mu_2 + \mu_3) \cdot \pi(B_1(1),2,1) = \mu_1 \cdot \pi(1,2,1)$$

$$\mu_2 \cdot \pi(B_1(2),2,0) = \mu_1 \cdot \pi(2,2,0) + \mu_3 \cdot \pi(B_1(1),2,1)$$

$$(\mu_1 + \mu_3) \cdot \pi(1,B_2(1),2) = \mu_2 \cdot \pi(1,1,2)$$

$$\mu_3 \cdot \pi(0,B_2(2),2) = \mu_2 \cdot \pi(0,2,2) + \mu_1 \cdot \pi(1,B_2(1),2)$$

$$(\mu_1 + \mu_2) \cdot \pi(2,1,B_3(1)) = \mu_3 \cdot \pi(2,1,1)$$

$$\mu_1 \cdot \pi(2,0,B_3(2)) = \mu_3 \cdot \pi(2,0,2) + \mu_2 \cdot \pi(2,1,B_3(1))$$

The global balance equations following from the Markov chain of Fig. 5.16 (Example 2) are presented below.

$$(\mu_{11} + \mu_2) \cdot \pi([1,1],2,0) = \mu_{12} \cdot \pi([1,2],1,B_3(1)) + \mu_3 \cdot \pi([0,1],2,1)$$

$$(\mu_{12} + \mu_2) \cdot \pi([1,2],2,0) = \mu_{11} \cdot \pi([1,1],2,0) + \mu_3 \cdot \pi([0,2],2,1)$$

$$(\mu_{12} + \mu_2 + \mu_3) \cdot \pi([1,2],1,1) = \mu_2 \cdot \pi([1,2],2,0) + \mu_3 \cdot \pi([0,2],1,2)$$
$$+ \mu_{11} \cdot \pi([1,1],1,1)$$

$$(\mu_{11} + \mu_2 + \mu_3) \cdot \pi([0,1],2,1) = \mu_2 \cdot \pi(B_1[1,2],2,0) + \mu_{12} \cdot \pi([1,2],1,1)$$
$$+ \mu_3 \cdot \pi([0,0],2,2)$$

$$(\mu_{12} + \mu_2 + \mu_3) \cdot \pi([0,2], 2, 1) = \mu_{11} \cdot \pi([0,1], 2, 1)$$

$$(\mu_{12} + \mu_2 + \mu_3) \cdot \pi([0,2], 1, 2) = \mu_2 \cdot \pi([0,2], 2, 1) + \mu_{11} \cdot \pi([0,1], 1, 2)$$

$$(\mu_{11} + \mu_2 + \mu_3) \cdot \pi([1,1], 1, 1) = \mu_2 \cdot \pi([1,1], 2, 0) + \mu_3 \cdot \pi([0,1], 1, 2)$$
$$+ \mu_{12} \cdot \pi([1,2], 0, B_3(2))$$

$$(\mu_{11} + \mu_3) \cdot \pi([1,1], 0, 2) = \mu_2 \cdot \pi([1,1], 1, 1) + \mu_3 \cdot \pi([0,1], B_2(1), 2)$$

$$(\mu_{12} + \mu_3) \cdot \pi([1,2], 0, 2) = \mu_{11} \cdot \pi([1,1], 0, 2) + \mu_2 \cdot \pi([1,2], 1, 1)$$
$$+ \mu_3 \cdot \pi([0,2], B_2(1), 2)$$

$$(\mu_{11} + \mu_2 + \mu_3) \cdot \pi([0,1], 1, 2) = \mu_{12} \cdot \pi([1,2], 0, 2) + \mu_2 \cdot \pi([0,1], 2, 1)$$
$$+ \mu_3 \cdot \pi([0,0], B_2(2), 2)$$

$$(\mu_2 + \mu_3) \cdot \pi([0,0], 2, 2) = \mu_{12} \cdot \pi([0,2], 1, 2) + \mu_2 \cdot \pi(B_1[0,2], 2, 1)$$

$$\mu_2 \cdot \pi(B_1[1,2], 2, 0) = \mu_{12} \cdot \pi([1,2], 2, 0) + \mu_3 \cdot \pi(B_1[0,2], 2, 1)$$

$$(\mu_2 + \mu_3) \cdot \pi(B_1[0,2], 2, 1) = \mu_{12} \cdot \pi([0,2], 2, 1)$$

$$\mu_3 \cdot \pi([0,0], B_2(2), 2) = \mu_{12} \cdot \pi([0,2], B_2(1), 2) + \mu_2 \cdot \pi([0,0], 2, 2)$$

$$(\mu_{11} + \mu_3) \cdot \pi([0,1], B_2(1), 2) = \mu_2 \cdot \pi([0,1], 1, 2)$$

$$(\mu_{12} + \mu_3) \cdot \pi([0,2], B_2(1), 2) = \mu_{11} \cdot \pi([0,1], B_2(1), 2) + \mu_2 \cdot \pi([0,2], 1, 2)$$

$$(\mu_{11} + \mu_2) \cdot \pi([1,1], 1, B_3(1)) = \mu_3 \cdot \pi([1,1], 1, 1)$$

$$\mu_{11} \cdot \pi([1,1], 0, B_3(2)) = \mu_2 \cdot \pi([1,1], 1, B_3(1)) + \mu_3 \cdot \pi([1,1], 0, 2)$$

$$(\mu_{12} + \mu_2) \cdot \pi([1,2], 1, B_3(1)) = \mu_3 \cdot \pi([1,2], 1, 1) + \mu_{11} \cdot \pi([1,1], 1, B_3(1))$$

$$\mu_{12} \cdot \pi([1,2], 0, B_3(2)) = \mu_2 \cdot \pi([1,2], 1, B_3(1))$$
$$+ \mu_{11} \cdot \pi([1,1], 0, B_3(2)) + \mu_3 \cdot \pi([1,2], 0, 2)$$

The set of equations corresponding to the Markov chain of Example 3, see Fig. 5.17, are provided subsequently.

$$(\mu_{11} + \mu_2) \cdot \pi([1,1], 2, 0) = \mu_{12} \cdot \pi([1,2], 1, B_3(1)) + \mu_3 \cdot \pi([0,1], 2, 1)$$
$$+ (1 - a) \cdot \mu_{11} \cdot \pi([1,1], 1, B_3(1))$$

$$(\mu_{12} + \mu_2) \cdot \pi([1,2], 2, 0) = a \cdot \mu_{11} \cdot \pi([1,1], 2, 0) + \mu_3 \cdot \pi([0,2], 2, 1)$$

$$(\mu_{12} + \mu_2 + \mu_3) \cdot \pi([1,2], 1, 1) = \mu_2 \cdot \pi([1,2], 2, 0) + \mu_3 \cdot \pi([0,2], 1, 2)$$
$$+ a \cdot \mu_{11} \cdot \pi([1,1], 1, 1)$$

$$(\mu_{11} + \mu_2 + \mu_3) \cdot \pi([0,1], 2, 1) = \mu_2 \cdot \pi(B_1[1,2], 2, 0) + \mu_{12} \cdot \pi([1,2], 1, 1)$$
$$+ \mu_3 \cdot \pi([0,0], 2, 2)$$

$$+ (1 - a) \cdot \mu_{11} \cdot \pi([1, 1], 1, 1)$$

$$+ \mu_2 \cdot \pi(B_1[1, 1], 2, 0)$$

$$(\mu_{12} + \mu_2 + \mu_3) \cdot \pi([0, 2], 2, 1) = a \cdot \mu_{11} \cdot \pi([0, 1], 2, 1)$$

$$(\mu_{12} + \mu_2 + \mu_3) \cdot \pi([0, 2], 1, 2) = \mu_2 \cdot \pi([0, 2], 2, 1) + a \cdot \mu_{11} \cdot \pi([0, 1], 1, 2)$$

$$(\mu_{11} + \mu_2 + \mu_3) \cdot \pi([1, 1], 1, 1) = \mu_2 \cdot \pi([1, 1], 2, 0) + \mu_3 \cdot \pi([0, 1], 1, 2)$$

$$+ \mu_{12} \cdot \pi([1, 2], 0, B_3(2))$$

$$+ (1 - a) \cdot \mu_{11} \cdot \pi([1, 1], 0, B_3(2))$$

$$(\mu_{11} + \mu_3) \cdot \pi([1, 1], 0, 2) = \mu_2 \cdot \pi([1, 1], 1, 1) + \mu_3 \cdot \pi([0, 1], B_2(1), 2)$$

$$(\mu_{12} + \mu_3) \cdot \pi([1, 2], 0, 2) = a \cdot \mu_{11} \cdot \pi([1, 1], 0, 2) + \mu_2 \cdot \pi([1, 2], 1, 1)$$

$$+ \mu_3 \cdot \pi([0, 2], B_2(1), 2)$$

$$(\mu_{11} + \mu_2 + \mu_3) \cdot \pi([0, 1], 1, 2) = \mu_{12} \cdot \pi([1, 2], 0, 2) + \mu_2 \cdot \pi([0, 1], 2, 1)$$

$$+ \mu_3 \cdot \pi([0, 0], B_2(2), 2)$$

$$+ (1 - a) \cdot \mu_{11}\pi([1, 1], 0, 2)$$

$$(\mu_2 + \mu_3) \cdot \pi([0, 0], 2, 2) = \mu_{12} \cdot \pi([0, 2], 1, 2) + \mu_2 \cdot \pi(B_1[0, 2], 2, 1)$$

$$+ \mu_2 \cdot \pi(B_1[0, 1], 2, 1)$$

$$+ (1 - a) \cdot \mu_{11} \cdot \pi([0, 1], 1, 2)$$

$$\mu_2 \cdot \pi(B_1[1, 2], 2, 0) = \mu_{12} \cdot \pi([1, 2], 2, 0) + \mu_3 \cdot \pi(B_1[0, 2], 2, 1)$$

$$(\mu_2 + \mu_3) \cdot \pi(B_1[0, 2], 2, 1) = \mu_{12} \cdot \pi([0, 2], 2, 1)$$

$$\mu_3 \cdot \pi([0, 0], B_2(2), 2) = \mu_{12} \cdot \pi([0, 2], B_2(1), 2) + \mu_2 \cdot \pi([0, 0], 2, 2)$$

$$+ (1 - a) \cdot \mu_{11} \cdot \pi([0, 1], B_2(1), 2)$$

$$(\mu_{11} + \mu_3) \cdot \pi([0, 1], B_2(1), 2) = \mu_2 \cdot \pi([0, 1], 1, 2)$$

$$(\mu_{12} + \mu_3) \cdot \pi([0, 2], B_2(1), 2) = a \cdot \mu_{11} \cdot \pi([0, 1], B_2(1), 2) + \mu_2 \cdot \pi([0, 2], 1, 2)$$

$$(\mu_{11} + \mu_2) \cdot \pi([1, 1], 1, B_3(1)) = \mu_3 \cdot \pi([1, 1], 1, 1)$$

$$\mu_{11} \cdot \pi([1, 1], 0, B_3(2)) = \mu_2 \cdot \pi([1, 1], 1, B_3(1)) + \mu_3 \cdot \pi([1, 1], 0, 2)$$

$$(\mu_{12} + \mu_2) \cdot \pi([1, 2], 1, B_3(1)) = \mu_3 \cdot \pi([1, 2], 1, 1) + a \cdot \mu_{11} \cdot \pi([1, 1], 1, B_3(1))$$

$$\mu_{12} \cdot \pi([1, 2], 0, B_3(2)) = \mu_2 \cdot \pi([1, 2], 1, B_3(1))$$

$$+ a \cdot \mu_{11} \cdot \pi([1, 1], 0, B_3(2)) + \mu_3 \cdot \pi([1, 2], 0, 2)$$

$$\mu_2 \cdot \pi(B_1[1, 1], 2, 0) = (1 - a) \cdot \mu_{11} \cdot \pi([1, 1], 2, 0)$$

$$+ \mu_3 \cdot \pi(B_1[0, 1], 2, 1)$$

$$(\mu_3 + \mu_2) \cdot \pi(B_1[0, 1], 2, 1) = (1 - a) \cdot \mu_{11} \cdot \pi([0, 1], 2, 1)$$

Algorithm 12 Construction of the transition rate matrix

1: **procedure** CONSTRUCTION OF THE TRANSITION RATE MATRIX
2:     **for each** $s \in \mathcal{S}$ **do**
3:         **for** $i = 1$ **to** M **do**
4:             **for** $nrt = 1$ **to** NRT_i **do**
5:                 Determine the target state t ▷ Sect. 5.5.2.1
6:                 Determine the transition rate μ^τ_{sti} and
7:                 insert it into $Q(s,t)$ and $Q(s,s)$ ▷ Sect. 5.5.2.2
8:             **end for**
9:         **end for**
10:     **end for**
11: **end procedure**

5.5.2 Construction of the Transition Rate Matrix

This section presents the construction of the transition rate matrix for the Markov-chain model of closed queueing networks. In order to determine the position of the transition rates in the transition rate matrix, the target states for all departures from all states are identified, see Sect. 5.5.2.1. The transition-rate values are assigned according to the kind of transition, see Sect. 5.5.2.2.

The transition rate matrix of the Markov-chain model is constructed according to Eqs. (5.88) and (5.89). The transition rate matrix, denoted by Q, has the dimension $|\mathcal{S}| \times |\mathcal{S}|$. It is composed of the transition rates of all originating states s to all target states t, where s and t are elements of the state space $\mathcal{S}$. The transition rates are denoted by $\mu^\tau_{sti} \; \forall s, t, i$, with i indicating the station inducing the transition.[49] The notation of this section is presented in Table 5.16.

Algorithm 12 provides an overview of the construction of the transition rate matrix. In order to determine the complete transition rate matrix, each transition possibility must be considered. On the first level, it is iterated over all states s. Within each s, it is iterated over all stations i and within that, over all nrt. nrt accounts for different exit probabilities from a phase. In each s, for each station i, in each nrt, it is investigated whether a transition can occur. If so, the parameters of the target state t are specified. The target state t is subject to the parameters of the origin state s, the station i inducing the change, and the exit nrt. This represents the first step in the construction of the transition rate matrix, see Sect. 5.5.2.1. In the second step, the value of the transition rate μ^τ_{sti} is identified and entered into the transition rate matrix at the positions $Q(s,t)$ and $Q(s,s)$, see Sect. 5.5.2.2.

---

[49] Although there is only one transition from state s to state t for each $s, t \in \mathcal{S}$ (i.e. there is only one station that can induce that particular transition), the transition rate μ^τ_{sti} further contains the information of i in order to assign its value, see Eq. (5.91) in Sect. 5.5.2.2.

5.5.2.1   Determination of the Target State t

For each state s, each station i and each exit nrt, it is checked whether a transition can potentially take place. If the conditions for a transition are fulfilled, the type of transition is detected. According to the type of transition and the values of s, i, and nrt, the parameters of t, namely $\vec{q}$, $\vec{ph}$, and $\vec{bs}$, are identified. In the following, ph, bs, and q contain the index of the considered state as an additional subscript.

The first condition for a transition from station i in state s constitutes that at least one customer is situated at station i. This is indicated by $ph_{si} \geq 1$, i.e. the process resides in the first or a higher phase. The second condition for a transition initiated by station i consists in the fact that the station must not be blocked in state s: $bs_{si} = 0$. In summary, the conditions for a transition from state s induced by station i are that

1. Station i contains at least one customer: $ph_{si} \geq 1$
2. Station i is not blocked: $bs_{si} = 0$.

If these conditions hold true, a transition occurs with rate μ_{sti}^{τ} from s to t caused by station i.

If the processing time follows a Coxian distribution and station i is active in the first phase, $ph_{si} = 1$, there are two possibilities for a workpiece to exit the station. For this reason, the algorithm iterates over nrt, the number of transition possibilities. The number of possible transitions is denoted by NRT_i. $nrt = 1$ indicates that the process will continue from phases 1 to 2, and $nrt = 2$ accounts for the possibility of exiting the station after the first phase. If there is no branching, the number of transitions equals one, $NRT_i = 1$.

The procedure for the determination of the target state t for given values of s, i, and nrt is illustrated in Fig. 5.20. The parameter values of the target state t are determined by adapting those of the origin state s according to the kind of transition. The type of transition is identified from the parameters of the origin state s (composed of queue lengths, indices of the active phases and blocking statuses), the station i inducing the change, and the branching type nrt.

We distinguish between a transition to the next phase (phase transition), a transition to the next station (station transition), a transition leading to blocking (blocking transition), and a transition resolving blocking (station transition and resolution of blocking).

A **phase transition** takes place if the process has not reached the last phase yet ($ph_{si} < ph_{si}^{NR}$) and if it is going to proceed to the next phase ($nrt = 1$). If station i induces the departure from state s in this case, the only change between the target state t and the origin state s is that the phase value is increased by 1: $ph_{ti} = ph_{si} + 1$.

If the process resides in the last phase ($ph_{si} = ph_i^{NR}$) or if the workpiece is leaving after the first phase ($nrt = 2$), further differentiations have to be made. It is distinguished whether station i will be blocked by the following state transition or not. If the subsequent buffer is full, i.e. $q_{s,i+1} = b_{i+1}$, the station will be blocked in the target state, so only the blocking status at station i is switched to "blocked": $bs_{ti} = 1$, a **blocking transition** occurs.

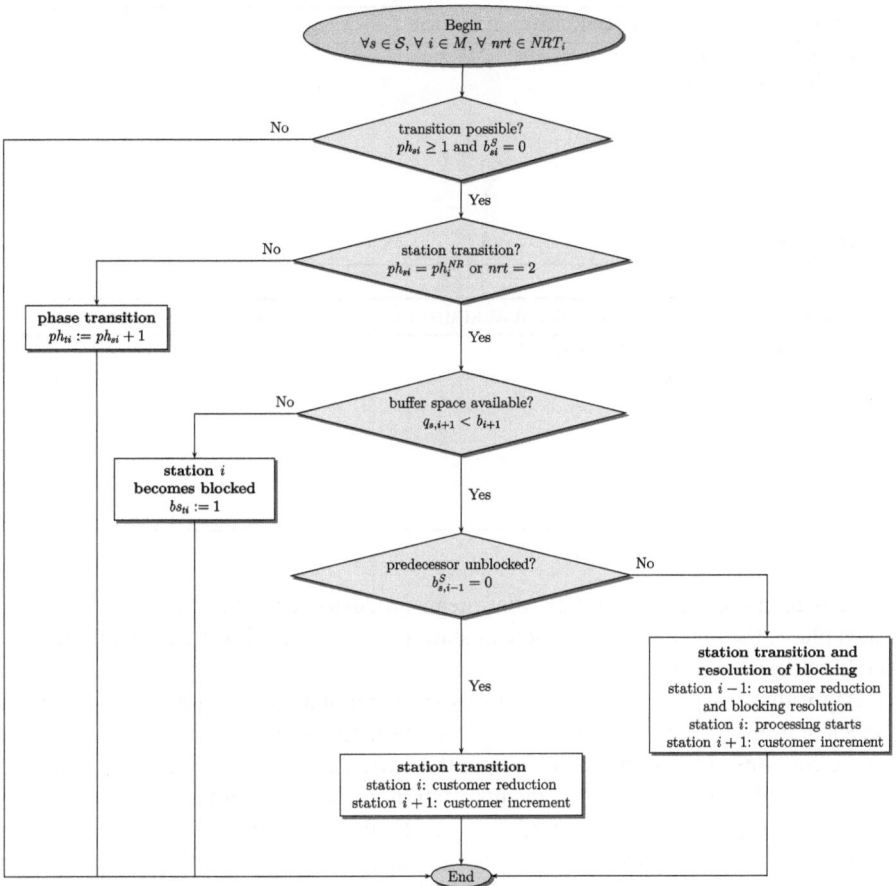

Fig. 5.20 Flowchart of the determination of the target state t

If the subsequent buffer has space available ($q_{s,i+1} < b_{i+1}$, yet with $ph_{si} = ph_{si}^{UB}$ or $nrt = 2$), a further distinction has to be made, namely whether the predecessor is blocked or not. If the predecessor station is not blocked ($bs_{s,i-1} = 0$), a customer leaves station i (customer reduction at station i) and station $i + 1$ receives it (customer increment at station $i + 1$). These changes constitute a **station transition**.

In the last case (still $ph_{si} = ph_{si}^{UB}$ or $nrt = 2$, and $q_{s,i+1} < b_{i+1}$, but $b_{s,i-1} = 1$) it is considered that the preceding station $i - 1$ is blocked in the origin state s. If station i is the first to finish its current active phase in this setting, a **station transition and resolution of blocking** occurs. Station i releases its current job, i.e. the customer departing from station i joins station $i + 1$ (customer increment at station $i + 1$) and station i starts processing on the next workpiece. A buffer space in front of station i becomes available and the workpiece formerly blocking station

---

Algorithm 13 Customer Reduction at station i

1: **procedure** CUSTOMER REDUCTION
2:     **if** $q_{si} = 0$ **then**
3:         $ph_{ti} := 0$
4:     **else**
5:         $q_{ti} := q_{si} - 1$
6:         $ph_{ti} := 1$
7:     **end if**
8: **end procedure**

Algorithm 14 Customer Increment at station $i + 1$

1: **procedure** CUSTOMER INCREMENT
2:     **if** $ph_{si} = 0$ **then**
3:         $ph_{ti} := 1$
4:     **else**
5:         $q_{ti} := q_{si} + 1$
6:     **end if**
7: **end procedure**

$i - 1$ is released into that buffer (that means, a customer reduction at station $i - 1$ takes place; the customer-movement at station i is neutral). Further, at station $i - 1$, the blockage is resolved, $bs_{t,i-1} := 0$.

Algorithm 13 presents the modifications from state s to state t if a customer leaves station i. The parameter values of the target state depend on whether the queue contains at least one workpiece in the buffer. If not, the phase value at station i is set equal to $ph_{ti} := 0$ (that means, station i is starving in the target state). If a workpiece is located in the buffer, $q_{si} \geq 1$, the first workpiece in the queue moves on to the server, and processing is started in phase 1. Hence, the phase value is set to $ph_{ti} = 1$ and the number of workpieces in the queue is reduced by 1, $q_{ti} := q_{si} - 1$.

Algorithm 14 shows how state t is adapted from state s if a customer arrives at station i. If no customer resides at station i ($ph_{si} = 0$), the arriving customer receives service immediately, namely in the first phase: $ph_{ti} := 1$. Otherwise, the queue length is increased by one, i.e. $q_{ti} := q_{si} + 1$.

In what follows, we investigate the state changes that arise if the blockage of station $i - 1$ is resolved. The corresponding pseudo code is shown in Algorithm 15. The actuator of the blocking resolution at station $i - 1$ is the workpiece-finish at station i. At that time-instant, the workpiece blocking the server of station $i - 1$ moves on to station i, and blocking at station $i - 1$ is resolved, i.e. $bs_{t,i-1} := 0$. One customer joins station $i + 1$ and a customer reduction at station $i - 1$ takes place. At station i, both an increment and a reduction take place, thus the workpiece level stays the same. However, station i starts with the next workpiece. Therefore, the phase value at station i is set to 1, $ph_{ti} := 1$. At a high number of n, additionally to the changes described above, it has to be considered that several stations could be blocked, all waiting for station i to finish its job. In this case, all workpieces which have blocked their server because of station $i - 1$ being blocked, move up

---

Algorithm 15 Resolution of blocking

1: **procedure** RESOLUTION OF BLOCKING
2:    CUSTOMER REDUCTION($i - 1$)
3:    CUSTOMER INCREMENT($i + 1$)
4:    $ph_{ti} := 1$
5:    $bs_{t,i-1} := 0$
6:    $j := i - 1$
7:    **repeat**
8:        CUSTOMER REDUCTION($j - 1$)
9:        CUSTOMER INCREMENT(j)
10:       $bs_{t,j-1} := 0$
11:       $j := j - 1$
12:   **until** $bs_{sj} = 0$
13: **end procedure**

---

one station at the time-instant of station i finishing its job. This means, there is a customer increment at station j and a customer reduction at the preceding station $j - 1$ for all blocked predecessor stations j. The blocking status at the upstream station, $j - 1$, is resolved as well: $bs_{t,j-1} := 0$. The phase value at station $j - 1$ depends on the queue length at that station and is taken into account in the function CUSTOMER REDUCTION($j - 1$).

5.5.2.2 Entries in the Transition Rate Matrix Q

Once the target state t is found, the transition rate μ^τ_{sti} is determined and entered in the transition rate matrix Q. The transition rate μ^τ_{sti} considers the transition from state s to state t induced by station i. The transition rate equals the rate of the current active phase at station i in state s, $\mu_{i,ph_{si}}$ multiplied by the probability of taking the exit nrt. The probability of taking branch $nrt = 1$ equals a_i and the probability in case of $nrt = 2$ is equal to $(1 - a_i)$. The value of a_i is specified by the phase-type distribution. If the processing time at station i is not Coxian distributed or if $ph_{si} > 1$, it holds that $a_i = 1$ and $nrt = 1$ (in our setting). The transition rates are assigned as shown in Eq. (5.91).

$$\mu^\tau_{sti} = \begin{cases} \mu_{i,ph_{si}} \cdot a_i & \text{if } nrt = 1 \\ \mu_{i,ph_{si}} \cdot (1 - a_i) & \text{if } nrt = 2 \end{cases} \tag{5.91}$$

In the first line of Eq. (5.91), it is considered that the process moves on to the next phase or finishes the last phase. The corresponding rate equals $\mu_{i,ph_{si}} \cdot a_i$. The second line of Eq. (5.91) accounts for the case that the processing time follows a Coxian distribution and that the process is exited after phase 1. This possibility is indicated by $nrt = 2$. It occurs with probability $(1 - a_i)$. The transition rate equals $\mu_{i,ph_{si}} \cdot (1 - a_i)$.

The transition rates μ^τ_{sti} are filled into the transition rate matrix Q. Q contains all transition rates from all states $s \in \mathcal{S}$ to all states $t \in \mathcal{S}$. It is of the dimension $|\mathcal{S}| \times |\mathcal{S}|$ and initialized by $Q(s,t) = 0 \; \forall s,t \in \mathcal{S}$. For each transition, the transition rate is considered as a departure from state s and as an arrival to state t.

The entries of the transition rate matrix follow from Eq. (5.88). The arrival to state t from state s is listed by the addition of the transition rate at the position $Q(s,t)$ of the matrix, see Eq. (5.92). The departure from state s is taken into account by subtracting the transition rate from $Q(s,s)$, see Eq. (5.93).

$$Q(s,t) := Q(s,t) + \mu^\tau_{sti} \qquad \forall s,t \in \mathcal{S} \tag{5.92}$$

$$Q(s,s) := Q(s,s) - \mu^\tau_{sti} \qquad \forall s,t \in \mathcal{S} \tag{5.93}$$

5.5.2.3 Example

This section provides the transition rate matrix of Example 1, see Table 5.17. It follows from the linear set of equations given in Sect. 5.5.1.3. The equations and Table 5.17 differ only by a rearrangement.

5.6 Solution of the Linear System of Equations

The solution of the linear set of equations, provided in Eqs. (5.88) and (5.90), yields the steady-state probabilities of the Markov chain, $\pi(s) \; \forall s \in \mathcal{S}$. In the following, the generalized minimal residual method (GMRES) is introduced. The GMRES procedure is very fast compared to other methods. Hence, it is the one used in the implementation.

5.6.1 GMRES

The Generalized Minimal Residual Method was proposed by Saad and Schultz (1986). It is an iterative numerical method serving to solve huge, sparse linear sets of equations. The benefit of this procedure is that it can handle non-symmetrical matrices. This is required in our case: Although the transition rate matrix Q is symmetrical, the incorporation of the normalizing constant into the system of equations leads to a matrix of the dimension $|\mathcal{S}| \times (|\mathcal{S}| + 1)$. For a comprehensive explanation of the GMRES procedure, we refer to Stewart (1994).[50]

---

[50]See Stewart (1994, pp. 190ff).

Table 5.17 Transition rate matrix for the Markov chain depicted in Fig. 5.15 (Example 1)

	$\pi(2,2,0)$	$\pi(2,1,1)$	$\pi(1,2,1)$	$\pi(1,1,2)$	$\pi(0,2,2)$	$\pi(2,0,2)$	$\pi(B_1(1),2,1)$	$\pi(B_1(2),2,0)$	$\pi(1,B_2(1),2)$	$\pi(0,B_2(2),2)$	$\pi(2,1,B_3(1))$	$\pi(2,0,B_3(2))$	
(1)	$-(\mu_1+\mu_2)$		μ_3					μ_1					$=0$
(2)	μ_2	$-(\mu_1+\mu_2+\mu_3)$		μ_3			μ_1						$=0$
(3)	μ_1		$-(\mu_1+\mu_2+\mu_3)$		μ_3				μ_2				$=0$
(4)		μ_1	μ_2	$-(\mu_1+\mu_2+\mu_3)$									$=0$
(5)			μ_1		$-(\mu_2+\mu_3)$								$=0$
(6)		μ_2				$-(\mu_1+\mu_3)$							$=0$
(7)				μ_1			$-(\mu_2+\mu_3)$						$=0$
(8)						μ_1		$-\mu_2$					$=0$
(9)				μ_2					$-(\mu_1+\mu_3)$				$=0$
(10)					μ_2					$-\mu_3$			$=0$
(11)											$-(\mu_1+\mu_2)$	μ_2	$=0$
(12)						μ_3					μ_2	$-\mu_1$	$=0$

5.6.2   Examples

In this section, Examples 1, 2, and 3 are continued to the solution of the steady-state probabilities of the Markov chain. The steady-state probabilities of Example 1 are provided first. Those of Examples 2 and 3 are presented consecutively. The workpiece levels n_i $\forall i$ of the corresponding states are listed next to the steady-state probabilities. n_i is obtained by summing up the workpieces on the server and in the buffer: $n_i = \mathbb{1}_{\{ph_i \geq 1\}} + q_i$.

Example 1.

$$\pi(2,2,0) = 0.121554653$$
$$\pi(2,1,1) = 0.08344242$$
$$\pi(1,2,1) = 0.113398126$$
$$\pi(1,1,2) = 0.056409461$$
$$\pi(0,2,2) = 0.048454396$$
$$\pi(2,0,2) = 0.035991482$$
$$\pi(B_1(1),2,1) = 0.056699063$$
$$\pi(B_1(2),2,0) = 0.272234828$$
$$\pi(1,B_2(1),2) = 0.017627956$$
$$\pi(0,B_2(2),2) = 0.040629742$$
$$\pi(2,1,B_3(1)) = 0.062581815$$
$$\pi(2,0,B_3(2)) = 0.090976059$$

Example 2.

	n_1	n_2	n_3
$\pi([0,0],2,2) = 0.044191$	0	2	2
$\pi([0,0],B_2(2),2) = 0.005009$	0	2	2
$\pi([0,1],1,2) = 0.050931$	1	1	2
$\pi([0,2],1,2) = 0.013122$	1	1	2
$\pi([0,1],B_2(1),2) = 0.014026$	1	1	2
$\pi([0,2],B_2(1),2) = 0.0411$	1	1	2
$\pi([0,1],2,1) = 0.102785$	1	2	1
$\pi([0,2],2,1) = 0.021518$	1	2	1
$\pi(B_1[0,2],2,1) = 0.045699$	1	2	1
$\pi([1,1],0,2) = 0.025813$	2	0	2
$\pi([1,2],0,2) = 0.009939$	2	0	2

$$\pi([1,1],0,B_3(2)) = 0.04354 \qquad 2 \quad 0 \quad 2$$
$$\pi([1,2],0,B_3(2)) = 0.020976 \qquad 2 \quad 0 \quad 2$$
$$\pi([1,1],1,1) = 0.068481 \qquad 2 \quad 1 \quad 1$$
$$\pi([1,1],1,B_3(1)) = 0.049153 \qquad 2 \quad 1 \quad 1$$
$$\pi([1,2],1,B_3(1)) = 0.016843 \qquad 2 \quad 1 \quad 1$$
$$\pi([1,2],1,1) = 0.02071 \qquad 2 \quad 1 \quad 1$$
$$\pi([1,1],2,0) = 0.100729 \qquad 2 \quad 2 \quad 0$$
$$\pi([1,2],2,0) = 0.032127 \qquad 2 \quad 2 \quad 0$$
$$\pi(B_1[1,2],2,0) = 0.273307 \qquad 2 \quad 2 \quad 0$$

Example 3.

	n_1	n_2	n_3
$\pi([0,0],2,2) = 0.05928140$	0	2	2
$\pi([0,0],B_2(2),2) = 0.04440098$	0	2	2
$\pi([0,1],1,2) = 0.03574449$	1	1	2
$\pi([0,2],1,2) = 0.00768465$	1	1	2
$\pi([0,1],B_2(1),2) = 0.00777054$	1	1	2
$\pi([0,2],B_2(1),2) = 0.00488989$	1	1	2
$\pi([0,1],2,1) = 0.08742035$	1	2	1
$\pi([0,2],2,1) = 0.01063508$	1	2	1
$\pi(B_1[0,1],2,1) = 0.07531227$	1	2	1
$\pi(B_1[0,2],2,1) = 0.00147300$	1	2	1
$\pi([1,1],0,2) = 0.01130403$	2	0	2
$\pi([1,2],0,2) = 0.01433573$	2	0	2
$\pi([1,1],0,B_3(2)) = 0.01369741$	2	0	2
$\pi([1,2],0,B_3(2)) = 0.15400541$	2	0	2
$\pi([1,1],1,1) = 0.03801158$	2	1	1
$\pi([1,2],1,1) = 0.01817824$	2	1	1
$\pi([1,1],1,B_3(1)) = 0.01800549$	2	1	1
$\pi([1,2],1,B_3(1)) = 0.02860876$	2	1	1
$\pi([1,1],2,0) = 0.05575897$	2	2	0
$\pi([1,2],2,0) = 0.02937511$	2	2	0
$\pi(B_1[1,1],2,0) = 0.27006322$	2	2	0
$\pi(B_1[1,2],2,0) = 0.01404341$	2	2	0

Table 5.18 Notation

d_i	Upper bound of the number of workpieces at station i, $d_i = \min\{n, b_i + 1\}$
D	Time between departures
L_i^Q	Mean queue length at station i
L_i^S	Mean work-in-process at station i
$P_i(n_i)$	Steady-state probability for n_i workpieces at station i
P_i^B	Blocking probability at station i
P_i^S	Probability of starvation at station i
PR	Production rate
U_i	Utilization at station i
T^S	Cycle time in the system
T_i^S	Cycle time at station i
T_i^B	Mean blocking time at station i
T_i^{Stv}	Mean starving time at station i
T_i	Mean processing time at station i, $T_i^P = 1/\mu_i$
T_i^Q	Mean waiting time in the buffer in front of station i

5.7 Performance Measures

The performance measures of the closed queueing networks are obtained in two
steps. First, the Markov-chain steady-state probabilities $\pi(s)$ $\forall s \in S$ are aggregated
to the CQN steady-state probabilities $P_i(n_i)$ and P_i^B, see Sect. 5.7.1. Second, the
CQN steady-state probabilities are used to calculate the performance measures, see
Sect. 5.7.2. Table 5.18 provides the notation used in this section.

5.7.1 Aggregation of the Markov-Chain Steady-State Probabilities

The steady-state probabilities per station, $P_i(n_i)$ $\forall i, n_i$, are obtained by summing
up all steady-state probabilities of the Markov chain containing the workpiece level
n_i at station i, disregarding the blocking status, see (5.94).

$$P_i(n_i) = \sum_{\substack{s \in S, \\ q_{si} + \mathbb{1}_{\{ph_{si} \geq 1\}} = n_i}} \pi(s) \qquad \text{for } n_i = 0, \ldots, \min\{b_i + 1, n\}, \forall i. \qquad (5.94)$$

$P_i(n_i)$ is calculated by the sum of all $\pi(s)$ over all states s for which it holds that
the queue length plus the workpiece on the server equals n_i. This is accounted for
by the condition $q_{si} + \mathbb{1}_{\{ph_{si} \geq 1\}} = n_i$. The parameter q_{si} denotes the queue length.
The term $\mathbb{1}_{\{ph_{si} \geq 1\}}$ equals 1 if in state s, station i is working (this means that the
value of the active phase is at least 1). If no workpiece is located on the server, the
term equals 0. The upper bound of n_i is either given by the overall workpieces in
the system, n, or by the station capacity, $b_i + 1$, whichever is less. Further, it holds

that the steady-state probabilities per station over all possible n_i must add up to 1:

$$\sum_{n_i=0}^{\min\{b_i+1,n\}} P_i(n_i) = 1 \qquad \forall i. \tag{5.95}$$

The blocking probability is obtained by summing up all steady-state probabilities of the Markov chain, $\pi_i(s)$, in which station i is blocked, regardless of the queue length, see Eq. (5.96).

$$P_i^B = \sum_{\substack{s \in \mathcal{S}, \\ bs_{si}=1}} \pi(s) \qquad \forall i. \tag{5.96}$$

5.7.2 Calculation of the Performance Measures

The performance measures of the closed queueing networks are obtained from the steady-state probabilities $P_i(n_i)$ for $0 \le n_i \le \min\{b_i + s_i, n\}$ workpieces at station i and the blocking probabilities P_i^B $\forall i$ as shown below.

The utilization at station i, denoted by U_i, conforms with the probability that station i is busy. Hence, it is calculated by the sum of the probabilities of one or more workpieces at the station minus the probability of blocking. Alternatively, it can be expressed as the probability of neither being starved nor blocked, see Eq. (5.97).

$$U_i = \sum_{n=1}^{d_i} P_i(n) - P_i^B = 1 - P_i^S - P_i^B \qquad \forall i \tag{5.97}$$

The parameter d_i denotes the maximum number of workpieces at station i, $d_i = \min\{n, b_i + s_i\}$ $\forall i$.

The blocking probability of station i, P_i^B, has to be derived from the performance analysis procedure and cannot be calculated from other performance measures.[51] The probability of starvation, denoted by P_i^S, can be obtained from the steady-state probabilities. It equals the probability that no workpiece resides at station i, see Eq. (5.98).

$$P_i^S = P_i(0) \qquad \forall i \tag{5.98}$$

The mean work-in-process at station i, denoted by L_i^S, results from the sum of workpieces weighted with its steady-state probabilities, see Eq. (5.99).

$$L_i^S = \sum_{n=1}^{c_i} n \cdot P_i(n) \qquad \forall i \tag{5.99}$$

It holds that $\sum_{i=1}^{M} L_i^S = n$.

---

[51] See Sect. 5.7.1.

The mean queue length or mean buffer level at station i, L_i^Q, equals the mean work-in-process at the entire station minus the work-in-process on the server. It is given in Eq. (5.100).

$$L_i^Q = L_i^S - \sum_{n_i=1}^{d_i} P_i(n_i) \qquad \forall i \tag{5.100}$$

The production rate PR is derived by multiplying the processing rate of station i by the utilization of station i, for any i. Due to the conservation of flow, the production rate is equal at all stations i.

$$PR = \mu_i \cdot U_i \qquad \forall i \tag{5.101}$$

The mean inter-departure time D corresponds to the reciprocal of the production rate, i.e $D = 1/PR$. Hence, it is also equal at all stations. It can alternatively be calculated by the sum of the processing time T_i, the mean blocking time, T_i^B, and the mean starving time, T_i^{Stv}, see Eq. (5.102). The determination of the complete inter-departure time distribution is presented in Chap. 6.

$$D = T_i + T_i^B + T_i^{Stv} \qquad \forall i \tag{5.102}$$

The mean blocking time is given by

$$T_i^B = P_i^B \cdot \frac{1}{PR} \qquad \forall i \tag{5.103}$$

and the mean starving time equals

$$T_i^{Stv} = P_i^S \cdot \frac{1}{PR} \qquad \forall i. \tag{5.104}$$

The mean blocking (starving) time is calculated by the probability of blocking (starving) P_i^B (P_i^S) multiplied by the mean time between departures at station i.

The mean cycle time at station i is denoted by T_i^S. According to Little's Law, it equals the ratio of the mean work-in-process L_i^S and the production rate PR, see Eq. (5.105).

$$T_i^S = \frac{L_i^S}{PR} \qquad \forall i \tag{5.105}$$

Little's Law also holds for the mean waiting time in the queue in front of station i, denoted by T_i^Q, see Eq. (5.106). T_i^Q is alternatively obtained from the cycle time in the system minus the time spent on the server. The time spent on the server equals the mean processing time plus the mean blocking time. It is also given in Eq. (5.106).

$$T_i^Q = \frac{L_i^Q}{PR} = T_i^S - \left(\frac{1}{\mu_i} + T_i^B \right) \qquad \forall i \tag{5.106}$$

Table 5.19 Steady-state probabilities of the exponential CQN (Example 1)

$P_i(n_i)$	$i = 1$	$i = 2$	$i = 3$
$n_i = 0$	0.089084137	0.126967541	0.393789481
$n_i = 1$	0.244134606	0.220061653	0.316121424
$n_i = 2$	0.666781257	0.652970807	0.290089095
P_i^B	0.32893389	0.058257698	0.153557874

Table 5.20 Performance measures of Example 1

Measure	not.	$i = 1$	$i = 2$	$i = 3$
Mean work-in-process	L_i^S	1.578	1.526	0.896
Blocking probability	P_i^B	0.329	0.0583	0.154
Utilization	ρ_i	0.582	0.815	0.453
Starving probability	P_i^S	0.089	0.127	0.394
Mean queue length	L_i^Q	0.667	0.653	0.29
Production rate	PR	0.407	0.407	0.407
Mean cycle time	T_i	3.873	3.746	2.2
Mean time blocked	T_i^B	0.807	0.143	0.377
Mean waiting time	T_i^Q	1.637	1.603	0.712

Table 5.21 Steady-state probabilities of the network with a hypo-exponential-2 distribution at station 1 (Example 2)

$P_i(n_i)$	$i = 1$	$i = 2$	$i = 3$
$n_i = 0$	0.085291	0.105881	0.406163
$n_i = 1$	0.25309	0.232662	0.319576
$n_i = 2$	0.661618	0.661456	0.27426
P_i^B	0.319006	0.060135	0.130512

The cycle time of the system, denoted by T^S, is determined by Little's Law as well, see Eq. (5.107).

$$T^S = \frac{n}{PR} \tag{5.107}$$

5.7.3   Examples

In the following, we present the steady-state probabilities and the performance measures for Examples 1, 2 and 3. The steady-state probabilities of Example 1 are displayed in Table 5.19. The corresponding performance measures are provided in Table 5.20. The steady-state probabilities of Example 2 are given in Table 5.21, and the resulting performance measures are shown in Table 5.22. The steady-state probabilities of Example 3 are listed in Table 5.23, and the corresponding performance measures are expressed in Table 5.24.

Table 5.22 Performance measures of Example 2

Measure	not.	$i = 1$	$i = 2$	$i = 3$
Mean work-in-process	L_i^S	1.576	1.556	0.868
Blocking probability	P_i^B	0.319	0.06	0.13
Utilization	ρ_i	0.596	0.834	0.463
Starving probability	P_i^S	0.085	0.106	0.406
Mean queue length	L_i^Q	0.662	0.661	0.274
Production rate	PR	0.417	0.417	0.417
Mean cycle time	T_i	3.78	3.73	2.081
Mean time blocked	T_i^B	0.765	0.144	0.313
Mean waiting time	T_i^Q	1.587	1.586	0.658

Table 5.23 Steady-state probabilities of the network with Cox-2 distribution at station 1 (Example 3)

$P_i(n_i)$	$i = 1$	$i = 2$	$i = 3$
$n_i = 0$	0.103682	0.193343	0.369241
$n_i = 1$	0.230930	0.158894	0.277645
$n_i = 2$	0.665387	0.647764	0.353115
P_i^B	0.360891903	0.057061406	0.214317065

Table 5.24 Performance measures of Example 3

Measure	not.	$i = 1$	$i = 2$	$i = 3$
Mean work-in-process	L_i^S	1.561	1.454	0.984
Blocking probability	P_i^B	0.36	0.057	0.214
Utilization	ρ_i	0.535	0.75	0.416
Starving probability	P_i^S	0.104	0.193	0.369
Mean queue length	L_i^Q	0.665	0.648	0.353
Production rate	PR	0.375	0.375	0.375
Mean cycle time	T_i	4.167	3.881	2.625
Mean time blocked	T_i^B	0.963	0.152	0.572
Mean waiting time	T_i^Q	1.775	1.728	0.942

5.8 Runtime Performance and Numerical Results

In this section, the runtime performance of the Markov-chain approach is investigated and numerical results are presented. The results refer to an Intel(R) Xeon(R) CPU with 2.40 GHz and 8 GB RAM using a Windows 7 (64 bit) operating system.

Our test set covers $M = \{2, 3, 4, 5, 6, 11\}$ stations and $b_i = \{1, 2, 3, 4, 5\}$ buffers at each station i. The coefficients of variation are $c_i^2 = \{1, 4, 0.64, 0.25, 0.125\}$. The phase-type distributions modeling these coefficients of variations are the exponential distribution (1 phase), the Cox-2 distribution (2 phases), the hypo-exponential-2 distribution (2 phases), the Erlang-4 distribution (4 phases), and the Cox-8 distribution (8 phases), respectively. All parameter values were combined. The number of stations, buffers, and phases were alternated in order to

Table 5.25 Development of the computation time for CQN with exponentially distributed processing times ($c_i^2 = 1 \; \forall i$) and n^*

M	$b_i \, \forall i$	n^*	$|\mathcal{S}|$	$PR(n^*)$	Computation time [h]	[min]	[s]	[ms]
2	1	3	4	0.558348083	0	0	0	1
2	2	4	5	0.594699580	0	0	0	<1
2	3	5	6	0.618579190	0	0	0	<1
2	4	6	7	0.635337847	0	0	0	<1
2	5	7	8	0.647651800	0	0	0	3
3	1	4	12	0.439327883	0	0	0	1
3	2	6	19	0.484367901	0	0	0	3
3	3	7	27	0.511631105	0	0	0	<1
3	4	9	37	0.532224284	0	0	0	15
3	5	10	48	0.546332043	0	0	0	31
4	1	6	40	0.416640043	0	0	0	15
4	2	8	81	0.465188476	0	0	0	62
4	3	10	142	0.497073565	0	0	0	156
4	4	12	227	0.519565432	0	0	0	359
4	5	14	340	0.536132743	0	0	0	703
5	1	7	140	0.418566317	0	0	0	125
5	2	10	351	0.467895699	0	0	0	578
5	3	12	750	0.500718977	0	0	1	937
5	4	15	1,401	0.523602832	0	0	5	<1
5	5	17	2,410	0.540148321	0	0	12	72
6	1	8	480	0.413330782	0	0	0	890
6	2	11	1,560	0.464864256	0	0	5	296
6	3	15	3,984	0.499041032	0	0	21	375
6	4	18	8,815	0.522632608	0	1	5	343
6	5	21	17,420	0.539665997	0	2	51	437
...								
11	1	15	259,248	0.401988549	5	34	27	828

investigate their influence on the computation time. The service rates were chosen to be $\mu_1, \ldots, \mu_{10} = \{0.7, 0.8, 0.6, 0.7, 0.9, 0.8, 0.75, 0.9, 0.8, 0.85\}$. These were not varied because they do not influence the runtime.

The computation time is composed of the construction of the transition rate matrix and the solution of the linear system of equations. The moment-matching and the calculation of the performance measures are carried out within a millisecond and are, therefore, negligible. The time for the construction of the transition rate matrix amounts to 1–5 % of the complete computation time. Most of the computation time is taken for the solution of the linear system of equations.

The computation time and the production rate (as the one selected performance measure) are displayed for the considered processing time distributions

Table 5.26 Development of the computation time for CQN with hypo-exponential-2 distributed processing times ($c_i^2 = 0.64 \ \forall i$) and n^*

M	$b_i \ \forall i$	n^*	$|\mathcal{S}|$	$PR(n^*)$	Computation time			
					(h)	(min)	(s)	[ms]
2	1	3	12	0.594327621	0	0	0	1
2	2	4	16	0.628438032	0	0	0	1
2	3	5	20	0.648812250	0	0	0	3
2	4	6	24	0.662105666	0	0	0	5
2	5	7	28	0.671291336	0	0	0	8
3	1	4	54	0.479544540	0	0	0	31
3	2	6	98	0.523335547	0	0	0	81
3	3	7	150	0.546701446	0	0	0	203
3	4	9	218	0.562876430	0	0	0	375
3	5	11	294	0.573015085	0	0	0	703
4	1	6	288	0.461708761	0	0	0	453
4	2	8	688	0.508514649	0	0	1	906
4	3	10	1,344	0.536516488	0	0	6	<1
4	4	12	2,320	0.554707152	0	0	15	859
4	5	14	3,680	0.567104716	0	0	32	156
5	1	7	1,530	0.464768167	0	0	5	812
5	2	10	4,862	0.512296715	0	0	35	281
5	3	12	11,930	0.540439385	0	2	11	562
5	4	15	24,822	0.558405938	0	6	28	281
5	5	17	45,930	0.570247371	0	15	56	93
6	1	9	7,884	0.460173292	0	0	54	750
6	2	12	34,984	0.510712594	0	8	8	671
6	3	15	108,452	0.539894020	0	37	34	515
6	4	18	271,584	0.558217360	3	13	34	296

in Table 5.25 (exponential), Table 5.26 (hypo-exponential-2), Table 5.27 (Cox-2), Table 5.28 (Erlang-4), and Table 5.29 (Cox-8), each for the production-rate maximizing number of workpieces n^*. In each case, the computable system size is restricted by the working memory.

Instances with exponentially distributed processing times with up to five-stations are solved within milliseconds. Six-station systems with five buffer spaces need about 2 min. The limit is reached for 11 stations each with one buffer space. For the hypo-exponential-2 distribution, the solution time amounts to milliseconds up to four-station systems each with one buffer space and three-station systems each with five buffer spaces. The limit is reached for six-station systems each with four buffer spaces. The computation time for Cox-2 distributed processing times is slightly higher. This difference is attributed to the fact that the transition rate matrix contains more entries and consists of several more states in the case of Cox-2 distributed processing times.

Table 5.27 Development of the computation time for CQN with Cox-2 distributed processing times ($c_i^2 = 4 \ \forall i$) and n^*

M	$b_i \forall i$	n^*	$|\mathcal{S}|$	$PR(n^*)$	(h)	(min)	(s)	(ms)

M	$b_i \forall i$	n^*	$|\mathcal{S}|$	$PR(n^*)$	Computation time (h)	(min)	(s)	(ms)
2	1	3	16	0.485310652	0	0	0	4
2	2	4	20	0.510419341	0	0	0	4
2	3	5	24	0.529408845	0	0	0	5
2	4	6	28	0.544754916	0	0	0	7
2	5	7	32	0.557645417	0	0	0	9
3	1	4	72	0.350163102	0	0	0	47
3	2	6	128	0.378894732	0	0	0	120
3	3	7	180	0.400057960	0	0	0	218
3	4	9	260	0.418413907	0	0	0	375
3	5	10	336	0.433518511	0	0	0	578
4	1	5	384	0.309163555	0	0	0	562
4	2	7	864	0.340831588	0	0	2	15
4	3	9	1,632	0.365783143	0	0	4	968
4	4	11	2,752	0.386507548	0	0	12	625
4	5	13	4,288	0.404230656	0	0	25	0
5	1	7	2,880	0.296252511	0	0	8	656
5	2	9	7,040	0.330184341	0	0	33	484
5	3	12	17,200	0.357727570	0	1	52	437
5	4	15	34,672	0.380287898	0	5	12	15
5	5	17	59,800	0.399840745	0	10	48	171
6	1	8	15,936	0.282381654	0	1	24	0
6	2	11	57,984	0.319110348	0	8	27	468
6	3	14	162,576	0.348183708	0	36	49	406
6	4	17	382,176	0.372269959	6	19	23	750
6	5	20	792,480	0.392759630	22	48	20	296

Instances with Erlang-4 distributed processing times are solved within milliseconds for two-station systems. The limit is reached for five-station systems each with three buffer spaces. For the Cox-8 distribution, the solution time amounts to seconds for two-station systems with three or more buffer spaces at each station. The limit is reached at four stations each with two buffer spaces.

Table 5.28 Development of the computation time for CQN with Erlang-4 distributed processing times ($c_i^2 = 0.25 \; \forall i$) and n^*

					Computation time			
M	$b_i \forall i$	n^*	$|\mathcal{S}|$	$PR(n^*)$	(h)	(min)	(s)	(ms)
2	1	3	40	0.652505231	0	0	0	19
2	2	4	56	0.676694790	0	0	0	48
2	3	5	72	0.687539989	0	0	0	67
2	4	6	88	0.693054309	0	0	0	111
2	5	7	104	0.696042318	0	0	0	163
3	1	4	300	0.548402964	0	0	0	643
3	2	6	604	0.579269953	0	0	2	603
3	3	7	972	0.589981669	0	0	7	78
3	4	9	1,468	0.595133596	0	0	14	984
3	5	10	2,028	0.597490658	0	0	30	93
4	1	6	2,800	0.542240425	0	0	19	843
4	2	8	7,536	0.574599712	0	1	53	453
4	3	10	15,856	0.588066852	0	6	36	703
4	4	12	28,784	0.594206598	0	17	50	750
4	5	14	47,344	0.597124358	0	39	4	281
5	1	7	24,500	0.546138400	0	6	51	468
5	2	10	93,924	0.577315854	0	54	30	359
5	3	12	253,140	0.589471422	8	18	14	187

Table 5.30 displays the computation time for instances from literature of open linear queueing systems as reported in Helber (2005).[52] As can be seen, the computation is very fast. Table 5.31 presents the performance measures for hypo-exponential-2 distributed processing times with $c^2 = 0.64$.

Systems with five or six stations appear in industry indeed. Also, small buffers are conform with the attempt to keep the work-in-process small. Decomposition approaches, for example, do not perform well in the case of small buffers. That means, the Markov-chain approach is useful for such systems because they can be analyzed exactly within a reasonable amount of time. Furthermore, the Markov

---

[52]The open queueing systems were modeled as a special case of closed queueing systems. The applicability of a CQN-method to a OQN-test-instance is given if the number of customers n is set higher than the overall capacity of buffers and servers excluding the buffer between the first and the last station with all other parameters kept equal, i.e.

$$n = 1 + M + \sum_{i=2}^{M} b_i \tag{5.108}$$

and $b_1 = n + 1$. That way, the first station can never be starved and the last can never be blocked.

Table 5.29 Development of the computation time for CQN with Cox-8 distributed processing times ($c_i^2 = 0.125 \ \forall i$) and n^*

M	$b_i \forall i$	n^*	$|\mathcal{S}|$	$PR(n^*)$	(h)	(min)	(s)	(ms)
					\multicolumn{4}{c}{Computation time}			
2	1	3	144	0.681134828	0	0	0	289
2	2	4	208	0.694231104	0	0	0	799
2	3	5	272	0.698090088	0	0	1	421
2	4	6	336	0.699351862	0	0	2	309
2	5	7	400	0.699778238	0	0	3	598
3	1	4	1,944	0.581428034	0	0	21	691
3	2	6	4,184	0.596096340	0	1	45	845
3	3	7	6,936	0.598928642	0	4	23	687
3	4	9	10,712	0.599700055	0	9	56	718
3	5	10	15,000	0.599913360	0	17	36	703
4	1	6	33,696	0.580169383	0	20	16	312
4	2	8	97,696	0.595393446	2	22	17	484

Table 5.30 Production rate of open queueing systems as reported in Helber (2005)

Inst.	M	μ_i	c_i^2	b_i	Helber	BLS	Altiok	Sim.	Lag.	t(ms)
T2 C2	3	0.5	0.5	1	0.380	0.384	–	0.382	0.380	240
T2 C3	3	0.5	0.8	1	0.349	0.347	–	0.351	0.351	50
T2 C4	3	0.5	2.0	1	0.298	0.272	–	0.296	0.294	80
T2 C7	3	0.5	0.6	1	0.443	0.441	–	0.427	0.426	50
T4 C1	3	0.5	0.75	2	0.386	0.385	0.368	0.385	0.387	90
T4 C2	3	0.5	2	2	0.327	0.303	0.338	0.322	0.322	212
T4 C3	3	0.5	2	2,9	0.360	0.345	0.368	0.360	0.358	788

References/Abbreviations:

Helber	Helber (2005)
BLS	Buzacott, Liu, and Shanthikumar (1995)
Altiok	Altiok (1996)
Sim.	Simulation
Lag.	this approach
(ms)	computation time in milliseconds of this approach

chain approach can be used as fine-tuning step instead of simulation after the approximate evaluation of a high number of configurations within an optimization procedure.

Table 5.31 Performance measures of instances with $c^2 = 0.64$ for $PR(n^*)$ at station $i = 1$, rounded to 4 fractional digits

M	b_i	n	PR	$E[U_1]$	P_1^B	T_1^B	T_1	L_1^S	L_1^Q	T_1^Q
2	1	3	0.5943	0.8490	0.2571	0.2540	2.6627	1.5825	0.5825	0.9801
2	2	4	0.6284	0.8978	0.2145	0.1627	3.5059	2.2033	1.2033	1.9147
2	3	5	0.6488	0.9269	0.1890	0.1127	4.4054	2.8583	1.8583	2.8642
2	4	6	0.6621	0.9459	0.1724	0.0818	5.3547	3.5453	2.5453	3.8443
2	5	7	0.6713	0.9590	0.1609	0.0611	6.3497	4.2625	3.2625	4.8600
3	1	4	0.4795	0.6851	0.2249	0.1627	2.4364	1.1684	0.4053	0.8452
3	2	6	0.5233	0.7476	0.2407	0.1266	3.1422	1.6444	0.8305	1.5870
3	3	7	0.5467	0.7810	0.2032	0.0613	3.4675	1.8957	1.0812	1.9776
3	4	9	0.5629	0.8041	0.2192	0.0535	4.0204	2.2630	1.4288	2.5383
3	5	11	0.5730	0.8186	0.2289	0.0488	4.5052	2.5816	1.7350	3.0278
4	1	6	0.4617	0.6596	0.3191	0.4888	3.1702	1.4637	0.5784	0.1728
4	2	8	0.5085	0.7264	0.2725	0.3239	3.6192	1.8404	0.9492	0.1659
4	3	10	0.5365	0.7665	0.2500	0.2461	4.0797	2.1888	1.2903	0.1578
4	4	12	0.5547	0.7924	0.2383	0.2024	4.5217	2.5082	1.6035	0.1517
4	5	14	0.5671	0.8101	0.2325	0.1754	4.9336	2.7979	1.8883	0.1474
5	1	7	0.4648	0.6640	0.3031	0.5444	3.3511	1.5575	0.6405	0.2231
5	2	10	0.5123	0.7319	0.2793	0.4288	4.5096	2.3102	1.3587	0.1935
5	3	12	0.5404	0.7721	0.2532	0.3039	5.0166	2.7112	1.7749	0.1925
5	4	15	0.5584	0.7977	0.2502	0.2914	6.2948	3.5150	2.5546	0.1756
5	5	17	0.5702	0.8146	0.2428	0.2520	6.7173	3.8305	2.8722	0.1707
6	1	9	0.4602	0.6574	0.3215	0.6189	3.6822	1.6945	0.7523	0.2004
6	2	12	0.5107	0.7296	0.2750	0.4425	4.7195	2.4103	1.4547	0.2045
6	3	15	0.5399	0.7713	0.2573	0.3651	5.8196	3.1420	2.1736	0.1935
6	4	18	0.5582	0.7975	0.2489	0.3226	6.9966	3.9056	2.9281	0.1830

Notation:

PR	Production rate
$E[U_i]$	Utilization at station i
P_i^B	Blocking probability at station i
T_i^B	Mean blocking time at station i
T_i	Cycle time at station i
L_i^S	Mean work-in-process at station i
L_i^Q	Mean queue length at station i
T_i^Q	Mean waiting time in the buffer in front of station i

Chapter 6
Distribution of the Time Between Processing Starts

The two previous chapters focused on the calculation of the first-moment performance measures. In the following, the calculation of the exact distribution of the time between processing starts, denoted by *TBPS*, is presented. This distribution concerns processes that take place at the instant of a processing-start.

As in the previous chapter, CQN with phase-type distributed processing times and finite buffer capacities are considered. To the best of our knowledge, there do not exist any procedures yet. In the approach proposed here, the distribution of *TBPS* is determined using the structure of the Markov-chain representation of CQN and its steady-state solution as input.[1] The distribution of *TBPS* results in a general phase-type distribution that is specified by its transition rate matrix.

The course of this chapter is as follows. First, the distribution of the time between processing starts is motivated. In Sect. 6.2, the structure of the *TBPS*-distribution is introduced. As preparation for the calculation of the components of the *TBPS*-distribution, the states of the Markov chain are assigned to subsets, see Sect. 6.3. In Sect. 6.4, the computation of the components is presented. The moments of the phase-type distribution are calculated by the matrix-geometric method of Neuts (1982) as shown in Sect. 6.5. Subsequently, two examples illustrate the concept in Sect. 6.6. Lastly, the influences of the configuration parameters on the first and the second moment of the *TBPS*-distribution are investigated, see Sect. 6.7.

---

[1] The calculation of these is described in Chap. 5.

S. Lagershausen, *Performance Analysis of Closed Queueing Networks*, Lecture Notes in Economics and Mathematical Systems 663, DOI 10.1007/978-3-642-32214-3_6, © Springer-Verlag Berlin Heidelberg 2013

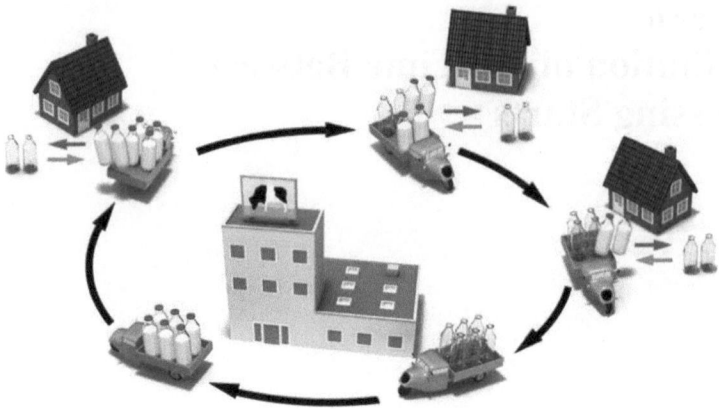

Fig. 6.1 Milk-run system

6.1   Motivation

The distribution of the time between processing starts is of interest for decisions concerning actions that take place at the instant of a processing-start. An example of such a decision is the dimensioning of buffers that are supplied by a so-called milk-run system.[2] It is illustrated in Fig. 6.1.[3]

The milk run is a concept to organize the delivery of parts or liquids that are transported in carriers. A supply vehicle approaches each station by a round trip. At the beginning of the round trip, the supply vehicle is loaded with full carriers. At each station, it picks up empty carriers and replaces them by full carriers. At the refill station, the empty carriers are dropped off and full carriers are loaded onto the supply vehicle. Thus, a new round trip can start.[4] The milk run is a method to speed up the time until a refill by routing vehicles that perform multiple pick-ups and drop-offs at many stations. Thereby, it is possible to reduce inventories and response times along a value stream.[5]

In the context of closed queueing networks, a milk run can be used to supply secondary buffers with material. A scheme of this is depicted in Fig. 6.2. The milk-run delivery is highlighted by the color red. The secondary buffers for the milk-run material are marked by the color blue. The standard CQN has the color black.

---

[2]See Werner (2010, p. 184f).

[3]See http://www.siegfeld.com/picflow/milkrun.jpg,03.04.2012

[4]See Werner (2010, p. 184f).

[5]See Lean Enterprise Institute (2003, p. 60).

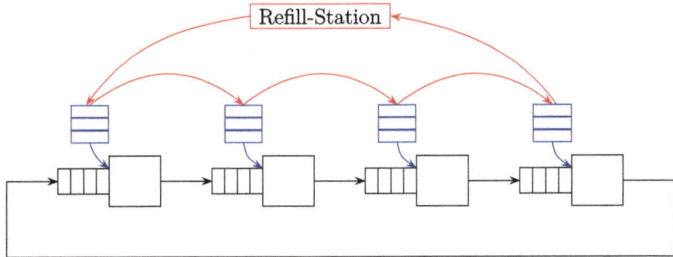

Fig. 6.2 Milk run in closed queueing systems

In such a system, an item from the secondary buffer is needed at the start-instant of the production process. In order to ensure the supply of material in the secondary buffers, the information on the distribution of the processing starts at each station is necessary. Also, in order to optimize the buffer sizes for the milk-run material, the coverage needs to be determined. This is done by a convolution of the probability density function of the time between processing starts. Therefore, the *TBPS*-distribution will be presented in the following.

6.2 Structure

The structure of the *TBPS*-distribution is derived from the time-course of the states at a certain station i. In the first time axis of Fig. 6.3, an exemplary course of states is depicted. The possible states are active, blocked, and starved. P_i denotes the processing time at station i. S_i stands for the starving time and B_i indicates the blocking time, each at station i. The notation is given in Table 6.1.

The time between two processing starts results from the realization of the processing, blocking, and starving sequence and its durations. It is depicted in the second time axis of Fig. 6.3. The processing time represents the first event of the time between two processing starts in any case. It is followed by blocking time, starving time, both, or nothing. The possible compositions of $TBPS_i$ are P_i, $P_i + S_i$, $P_i + B_i$, and $P_i + B_i + S_i$.

Note that the expected value of the time between certain events is equal, no matter if the event is the processing start, the processing end, or the departure of a workpiece. The reciprocal of the expected inter-event time equals the production rate. However, higher-order moments of these inter-event times differ.

The blocking and starving states are, in the following, subsumed into non-active states. The time between two processing starts at station i is composed of the processing time P_i and with the probability of transiting into non-activity, denoted by $\pi_{\mathcal{N}_i}$, additionally of non-active time, denoted by N_i. With the probability $1 - \pi_{\mathcal{N}_i}$, the service of the next workpiece is started after a workpiece-completion. Hence, the

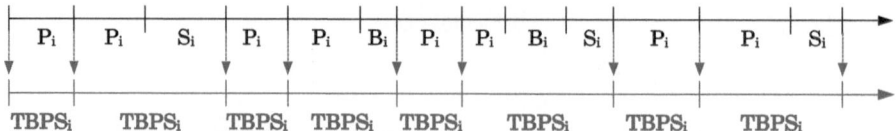

Fig. 6.3 Exemplary course of states

Table 6.1 Notation

B_i	Blocking time at station i
N_i	Non-active time at station i
P_i	Processing time at station i
S_i	Starving time at station i
$TBPS_i$	Time between two processing starts at station i

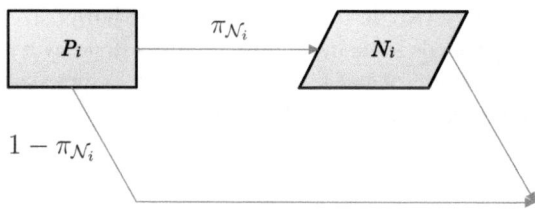

Fig. 6.4 Scheme of the time between processing starts

time between two processing starts equals $TBPS_i = P_i + \pi_{\mathcal{N}_i} \cdot N_i$. A scheme of $TBPS_i$ as described above is depicted in Fig. 6.4.

The structure of $TBPS_i$ is presented in full detail in Fig. 6.5. The first phase constitutes the processing time distribution (here: exponential distribution). The two subsequent squares represent hypothetical decision nodes. These do not take up time. The first square, e_P, marks the end of the processing time. With probability $\pi_{\mathcal{N}_i}$, the station enters non-activity, and with probability $(1 - \pi_{\mathcal{N}_i})$, $TBPS_i$ is finished, i.e. the next processing is started. The second square, s_N, indicates the start of the non-active time. In this decision node, it is determined which particular non-active state is entered. The probability of a transition from s_N into the particular non-active state u is denoted by $\pi_{T_i}(u)$. The non-active state u finishes with the exponential rate $\lambda_{\mathcal{N}_i}(u)$. The probability of transiting between two non-active states u and v is denoted by $p_N(u, v)$. The last node represents the absorbing state, labeled by as. The notation is displayed in Table 6.2.

The transition probabilities and transition rates displayed in Fig. 6.5 are needed for the calculation of the distribution of $TBPS_i$. As preparatory work for these calculations, the states of the Markov-chain representation of the CQN, $\mathcal{S}$, are assigned to different subsets.

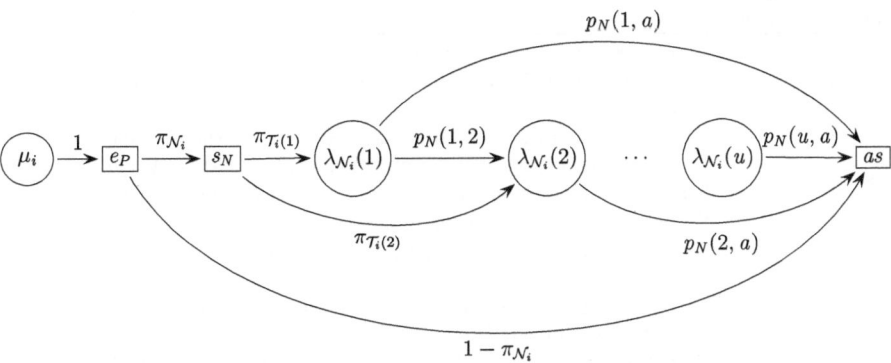

Fig. 6.5 Detailed scheme of the time between processing starts

Table 6.2 Notation

as	Absorbing state
e_P	Decision node after end of processing
$\lambda_{\mathcal{N}_i}(u)$	Exponential rate of non-active state u regarding station i
$\pi_{\mathcal{N}_i}$	Probability of transition from activity to non-activity regarding station i
$\pi_{\mathcal{T}_i}(ph, u)$	Probability of transiting into a particular non-active state u from an active state regarding station i
$\pi_{\mathcal{N}\mathcal{T}_i}(u)$	Probability of transiting into a particular non-active state u after the finish of a workpiece at station i
$p_N(u, v)$	Transition probability from non-active state u to non-active state v
s_N	Decision node at the start of the non-active time

6.3   Assignment of the Markov-Chain States to Subsets

The assignment of the states of the Markov-chain representation, $\mathcal{S}$, depends on the activity status, the transition possibilities into active or non-active states, and the current active phase. The notation of the different sets is summarized in Table 6.3.

The states are first distinguished between active and non-active states with regard to station i. The set of non-active states concerning station i is denoted by $\mathcal{N}_i$. As mentioned above, non-activity is present in the case of starving or blocking. A station is starved in state s if the current phase value equals zero, $ph_{si} = 0$. It is blocked if the blocking status equals 1, $bs_{si} = 1$. All states of the state set $\mathcal{S}$ in which station i is either starved or blocked belong to the subset $\mathcal{N}_i$, see Eq. (6.1).

$$\mathcal{N}_i = \{\forall s \in \mathcal{S} \mid ph_{si} = 0 \vee bs_{si} = 1\} \quad \forall i \in M \tag{6.1}$$

All states in which station i is active are collected in the set $\mathcal{A}_i$. Station i is active if the process at station i is situated at least in phase 1 and if it is not blocked, i.e. $ph_{si} \geq 1$ and $bs_{si} = 0$, see Eq. (6.2).

Table 6.3 Notation

$\mathcal{A}_i$	Set of processing states regarding station i
$\mathcal{A}_i^F(ph)$	Set of active states from which the workpiece at station i is potentially finished in the ph-th phase, $\mathcal{A}_i^F(ph) \subseteq \mathcal{A}_i$
$\mathcal{A}_i^{FN}(ph)$	Set of potential transition-into-non-activity states in which station i potentially finishes after the ph-th phase, $\mathcal{A}_i^{FN}(ph) \subseteq \mathcal{A}_i^F(ph)$
$\mathcal{P}_i^E$	Set of exits of the phase-type processing time distribution at station i
$\mathcal{N}_i$	Set of non-active states regarding station i with $\mathcal{N}_i = \{\mathcal{N}_i^E, \mathcal{N}_i^N\}$
$\mathcal{N}_i^E$	Set of non-active states that can be entered from an active state of station i, $\mathcal{N}_i^E \subseteq \mathcal{N}_i$
$\mathcal{N}_i^N$	Set of non-active states which can only be entered by another non-active state regarding station i, $\mathcal{N}_i^N \subseteq \mathcal{N}_i$
Q	Transition rate matrix of the Markov chain representing the CQN
S	State set of the Markov chain representing the CQN, with $S = \{\mathcal{A}_i, \mathcal{N}_i\}$

$$\mathcal{A}_i = \{\forall s \in S \mid ph_{si} \geq 1 \wedge bs_{si} = 0\} \quad \forall i \in M \tag{6.2}$$

A subset of $\mathcal{A}_i$ is the set $\mathcal{A}_i^F(ph)$. This set unites all those states in which station i is active in phase ph and from which the workpiece at station i can be finished by a transition out of that state (induced by station i). Therefore, for the states of $\mathcal{A}_i^F(ph)$, it is required that the process can be exited after phase ph. The indices of all exit phases of the processing time distribution at station i are collected in the set $\mathcal{P}_i^E$. So, all those states belong to the set $\mathcal{A}_i^F(ph)$ in which station i is active in phase ph, where ph represents an exit phase, $ph_{si} = ph$, $ph \in \mathcal{P}_i^E$. The set $\mathcal{A}_i^F(ph)$ is given in Eq. (6.3).

$$\mathcal{A}_i^F(ph) = \{s \in S \mid ph_{si} = ph, \ ph_{si} \in \mathcal{P}_i^E, \ s \in \mathcal{A}_i\}$$

$$\forall i \in M, \forall ph \in \mathcal{P}_i^E \tag{6.3}$$

From the set $\mathcal{A}_i^F(ph)$, another subset is defined, denoted by $\mathcal{A}_i^{FN}(ph)$. It unites all states that lead into non-activity if station i induces the state change by finishing phase ph. In other words, a state s belongs to the set $\mathcal{A}_i^{FN}(ph)$ if s is an element of $\mathcal{A}_i^F(ph)$ and if further, as station i finishes phase ph, it transits into non-activity. Such a transition exists if the entry in the transition rate matrix from state s to any state t is positive, where t belongs to the set of non-active states regarding station i, $Q(s, t) > 0$ with $s \in \mathcal{A}_i^F(ph)$ and any $t \in \mathcal{N}_i$.

$$\mathcal{A}_i^{FN}(ph) = \{s \in S \mid Q(s, t) > 0, \ s \in \mathcal{A}_i^F(ph), t \in \mathcal{N}_i\}$$

$$\forall i \in M, \forall ph \in \mathcal{P}_i^E \tag{6.4}$$

The set of non-active states $\mathcal{N}_i$ is also further divided. The subset of non-active states (regarding station i) that can be entered from an active state (concerning station i) is denoted by $\mathcal{N}_i^E$. These are the so-called non-active entrance states. State s is an element of the non-active entrance states $\mathcal{N}_i^E$ if there is a transition from any active state r to the non-active state s, i.e. $Q(r, s) > 0$ with any $r \in \mathcal{A}_i$ (more precisely (it falls together), $r \in \mathcal{A}_i^{FN}(ph)$ of any ph) and $s \in \mathcal{N}_i$, see Eq. (6.5).

$$\mathcal{N}_i^E = \{s \in \mathcal{S} \mid Q(r, s) > 0, \ s \in \mathcal{N}_i, r \in \mathcal{A}_i\} \quad \forall i \in M \qquad (6.5)$$

The complementary set of $\mathcal{N}_i^E$ in regard to the non-active states constitutes the set $\mathcal{N}_i^N$. Hence, it holds that $\mathcal{N}_i = \{\mathcal{N}_i^E, \mathcal{N}_i^N\}$. $\mathcal{N}_i^N$ represents the set of non-active states which can only be entered from another non-active state of station i. A state s is part of the subset $\mathcal{N}_i^N$ if there is any non-active state r with a transition into the non-active state s, i.e. $Q(r, s) > 0$ with $r, s \in \mathcal{N}_i$, see Eq. (6.6).

$$\mathcal{N}_i^N = \{s \in \mathcal{S} \mid Q(r, s) > 0, \ s \in \mathcal{N}_i, r \in \mathcal{N}_i\} \quad \forall i \in M \qquad (6.6)$$

6.4   Components of the Distribution

The processing time distribution is the only element of the $TBPS_i$-distribution that is pre-specified. All other components have to be calculated. The states, the transitions, and the steady-state probabilities of the continuous-time Markov chain of the CQN are required for these calculations. These are obtained as described in Chap. 5.

The probabilities of transition into non-active states are calculated as shown in Sect. 6.4.1. The exponential rates of non-active states u, $\lambda_{\mathcal{N}_i}(u)$, and the transition probabilities between non-active states u and v, $p_N(u, v)$, are specified in Sects. 6.4.2 and 6.4.3, respectively.

6.4.1   Probabilities of Transition into Non-active States

The probability of transition into non-activity, $\pi_{\mathcal{N}_i}$, and the probability of transition into a particular non-active state u from phase ph, $\pi_{\mathcal{T}_i}(ph, u)$, are calculated using the defined subsets in Sects. 6.4.1.1 and 6.4.1.2, respectively. From these expressions, the direct probability of transiting from phase ph to non-active state u, denoted by $\pi_{\mathcal{N}\mathcal{T}_i}(ph, u)$, is derived, see Sect. 6.4.1.3. The transition probability $\pi_{\mathcal{N}\mathcal{T}_i}(ph, u)$ is applied to build the transition rate matrix of the $TBPS_i$-distribution.

6.4.1.1   Probability of Transition into Non-activity

The probability of transiting into non-activity at station i after finishing phase ph is denoted by $\pi_{\mathcal{N}_i}(ph)$ for all $ph \in \mathcal{P}_i^E$. As depicted in Fig. 6.5, this decision is made in the hypothetical node e_p. The point in time when reaching this node corresponds in the Markov chain to the time-instance in which station i finishes the current workpiece after phase ph. Hence, at this point in time, it is taken for granted that station i induces the transition from state s.

As stated above, $TBPS_i$ starts with the processing time. After finishing a workpiece, station i either starts working on the next workpiece or it transits into non-activity. In the Markov chain, the event of process-ending can occur from any state s in which station i is active and may exit after the current phase ph, $s \in \mathcal{A}_i^F(ph)$ for all $ph \in \mathcal{P}_i^E$. If in the target state of the transition, station i is blocked or starved, $TBPS_i$ proceeds with non-active time. Otherwise, $TBPS_i$ is finished. $\pi_{\mathcal{N}_i}(ph)$ can be stated as the probability that the system resides in a state that transits into non-activity (event B) given that station i is active in phase ph with $ph \in \mathcal{P}_i^E$ (event A), see Eq. (6.7).

$$\pi_{\mathcal{N}_i}(ph) = P(B|A) = \frac{P(A \cap B)}{P(A)} \tag{6.7}$$

The states in which station i can exit the processing after phase ph (event A) are summarized in the set $\mathcal{A}_i^F(ph)$. The probability of residing in state $s \in \mathcal{A}_i^F(ph)$ equals the sum of the steady-state probabilities of all states that are an element of the state set $s \in \mathcal{A}_i^F(ph)$, see Eq. (6.8).

$$\begin{aligned} P(A) &= P(\text{residence in any } s \in \mathcal{A}_i^F(ph)) \\ &= \sum_{s \in \mathcal{A}_i^F(ph)} \pi(s) \qquad \forall i \in M, \forall ph \in \mathcal{P}_i^E \end{aligned} \tag{6.8}$$

The residence in a state that leads into non-activity regarding station i (event B) and in which station i is active in exit-phase ph (event A) applies to all states t that are an element of the set $A_i^{FN}(ph)$. The probability of transiting into non-activity from phase ph is given in Eq. (6.9).

$$\begin{aligned} P(A \cap B) &= P(\text{residence in any } t \in \mathcal{A}_i^{FN}(ph)) \\ &= \sum_{t \in \mathcal{A}_i^{FN}(ph)} \pi(t) \qquad \forall i \in M, \forall ph \in \mathcal{P}_i^E \end{aligned} \tag{6.9}$$

We can, therefore, state the probability $\pi_{\mathcal{N}_i}(ph)$ by inserting Equations (6.8) and (6.9) into (6.7). Hence, $\pi_{\mathcal{N}_i}(ph)$ is expressed as the probability of residing in any finishing-after-phase-ph-and-transiting-into-non-activity state t (i.e. any $t \in \mathcal{A}_i^{FN}(ph)$) based on the probability of residing in any finishing-after-phase-ph state s (i.e. any $s \in \mathcal{A}_i^F(ph)$), see Eq. (6.10).

$$\pi_{\mathcal{N}_i}(ph) = \frac{\sum_{t \in \mathcal{A}_i^{FN}(ph)} \pi(t)}{\sum_{s \in \mathcal{A}_i^F(ph)} \pi(s)} \qquad \forall i \in M, \forall ph \in \mathcal{P}_i^E \tag{6.10}$$

6.4.1.2 Probability of Transition into Particular Non-active States Given that Non-activity Is Entered

Given that station i enters non-activity, it must be determined which particular non-active state is entered. The probability that the system enters non-active state u when exiting after phase ph into non-activity is denoted by $\pi_{T_i}(ph, u)$.

As the model of the $TBPS_i$-distribution in Fig. 6.5 implies, the choice of the non-active entrance state is made in the hypothetical node s_N. At this point in time, it is given that station i has just finished its current workpiece after phase ph and that in the following, it will enter non-activity. The particular non-active state u is entered with probability $\pi_{T_i}(ph, u)$. It equals the probability of transiting into the non-active state u (event C) given that the system resides in any state in which station i is active in the process-ending phase ph (event $A \cap B$), that means

$$\pi_{T_i}(ph, u) = P(C | A \cap B) = \frac{P(A \cap B \cap C)}{P(A \cap B)}. \tag{6.11}$$

Let us consider $P(A \cap B \cap C)$. Therefore, we introduce the state $afn(ph, u)$. It denotes the active state which leads into the non-active state u when station i finishes its workpiece after phase ph.[6] Hence, state $afn(ph, u)$ is an element of the state set $\mathcal{A}_i^{FN}(ph)$. The probability of residing in state $afn(ph, u)$ is equal to the steady-state probability of that state, $P(A \cap B \cap C) = \pi(afn(ph, u))$.

According to Eq. (6.11), $\pi_{T_i}(ph, u)$ can be expressed as the probability of residing in state $afn(ph, u)$ based on the probability of residing in any state s that is active in phase ph and that can transit into non-activity, $s \in \mathcal{A}_i^{FN}(ph)$. $\pi_{T_i}(ph, u)$ holds for all exit phases, $ph \in \mathcal{P}_i^E$ and all non-active entrance states, $u \in \mathcal{N}_i^E$ and is provided in Eq. (6.12).

$$\pi_{T_i}(ph, u) = P(afn(ph, u) \mid \text{any } t \in \mathcal{A}_i^{FN}(ph))$$

$$= \frac{\pi(afn(ph, u))}{\sum\limits_{t \in \mathcal{A}_i^{FN}(ph)} \pi(t)} \quad \forall i \in M, \ \forall u \in \mathcal{N}_i^E, \forall ph \in \mathcal{P}_i^E \tag{6.12}$$

6.4.1.3 Probability of Transition into the Particular Non-active State u When Ending the Processing After Phase ph

The transition probabilities $\pi_{\mathcal{N}_i}(ph)$ and $\pi_{T_i}(ph, u)$ introduced above are now used to express the transition probability from phase ph to non-active state u directly.

---

[6]Note that this assignment is unambiguous. A non-active state might have several active predecessors, however, only one predecessor from an active state can finish after phase ph. This is because for any given state s, there is only one transition possibility to enter the particular non-active state u when requiring that the transition is induced by station i.

$\pi_{\mathcal{N}T_i}(ph, u)$ indicates the probability that station i enters non-active state u after the process completion after phase ph. This event happens (1) if the system transits into non-activity at all, given that process is currently in the active phase ph (its probability equals $\pi_{\mathcal{N}_i}(ph)$) and (2) given that it transits into non-activity from phase ph, that it transits into non-active state u (provided by $\pi_{T_i}(ph, u)$). Since these two events are independent, $\pi_{\mathcal{N}T_i}(ph, u)$ is calculated by the product of the corresponding probabilities, see Eq. (6.13).

$$\pi_{\mathcal{N}T_i}(ph, u) = \pi_{\mathcal{N}_i}(ph) \cdot \pi_{T_i}(u)$$

$$= \frac{\sum\limits_{t \in \mathcal{A}_i^{FN}(ph)} \pi(t)}{\sum\limits_{s \in \mathcal{A}_i^{F}(ph)} \pi(s)} \cdot \frac{\pi(afn(ph, u))}{\sum\limits_{t \in \mathcal{A}_i^{FN}(ph)} \pi(t)}$$

$$= \frac{\pi(afn(ph, u))}{\sum\limits_{s \in \mathcal{A}_i^{F}(ph)} \pi(s)} \qquad \forall i \in M, \forall ph \in \mathcal{P}_i^{E}, \forall u \in \mathcal{N}_i^{E} \quad (6.13)$$

As can be seen from Eq. (6.13), $\pi_{\mathcal{N}T_i}(ph, u)$ results in the probability of residing in the active state $afn(ph, u)$ based on the probability of residing in any residing-in-and-finishing-after-phase-ph state s.

Note that $\pi_{\mathcal{N}T_i}(ph, u) = 0$ for all non-active states u that cannot be entered by an active state, i.e. $\forall u \in \mathcal{N}_i^{N}$. This holds for all exit phases, $ph \in \mathcal{P}_i^{E}$. Also, $\pi_{\mathcal{N}T_i}(ph, u) = 0$ for all phases which are no exit phases, $ph \notin \mathcal{P}_i^{E}$, for all non-active entrance-states $u \in \mathcal{N}_i^{E}$.

Using the transition probabilities $\pi_{\mathcal{N}T_i}(ph, u)$, the model of Fig. 6.5 can be transferred into the model shown in Fig. 6.6. In that figure, the hypothetical decision nodes are eliminated. It represents the $TBPS_i$-distribution for the exponential processing time distribution. The $TBPS_i$-distribution for a processing time distribution with two phases and branching (Cox-2 distribution) is depicted in Fig. 6.7. The first two phases are the processing phases, the others are non-active states.

Apart from the probabilities of entrance into non-activity, the time spent in non-active states and the transition probabilities between these states are needed for the $TBPS_i$-distribution. These are introduced in the following.

6.4.2   Time Spent in a Non-active State

In the following, the time spent in a non-active state is determined. Each state endures until the first of all active phases at any station ends. That means, the minimum of the exponential processing phases over all stations that are active in a

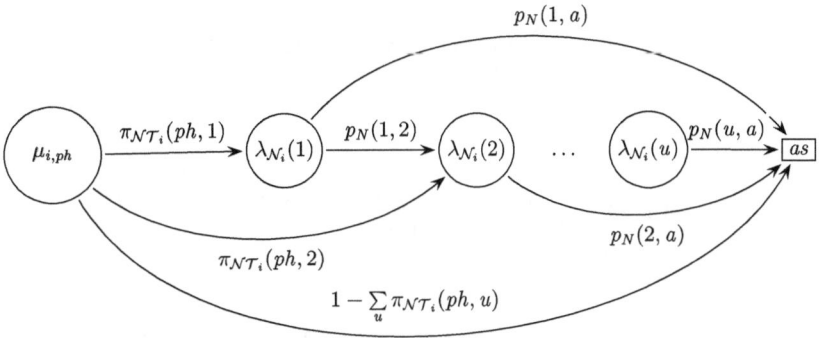

Fig. 6.6 *TBPS_i*-distribution with exponentially distributed processing times

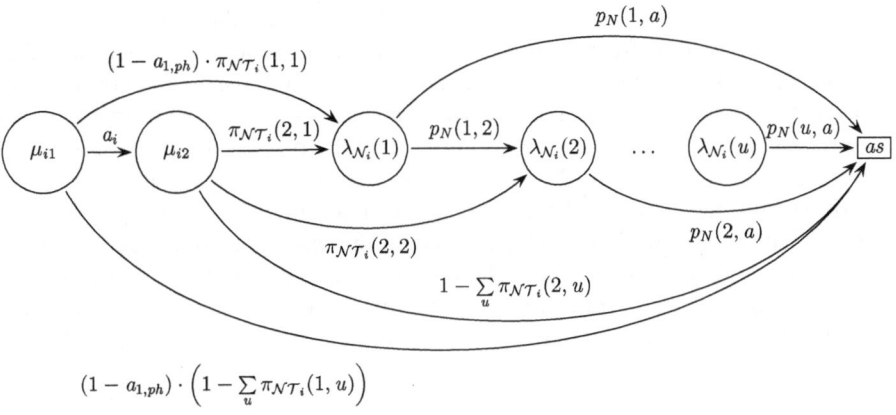

Fig. 6.7 *TBPS_i*-distribution with Cox-2 distributed processing times

state is needed.[7] According to Property 5.2,[8] the minimum of several exponentially distributed random variables is also exponentially distributed, namely with the sum of all rates on which the minimum is applied. Hence, the time in non-active state u is exponentially distributed as well with the rate $\lambda_{\mathcal{N}_i}(u)$.

The rate of non-active state u, $\lambda_{\mathcal{N}_i}(u)$, is equal to the sum of all phases that are active in state u. A station is active if its process resides at least in the first phase and if the station is not blocked. When station j is active in phase ph, it finishes that state with rate $\mu_{j,ph_{u,j}}$. If station j is not active, its rate amounts to zero. The summation of the rates of the active phases in state u over all stations j yields the rate of non-active state u, $\lambda_{\mathcal{N}_i}(u)$, see Eqs. (6.14) and (6.15).

---

[7]See Property 5.3 on page 94.
[8]See page 93.

$$\lambda_{\mathcal{N}_i}(u) = \sum_{j=1}^{M} \delta(u, j) \qquad \forall i, j \in M, \ \forall u \in \mathcal{N}_i \qquad (6.14)$$

with

$$\delta(u, j) = \begin{cases} \mu_{j, ph_{u,j}} & \text{if } ph_{u,j} \geq 1 \text{ and } bs_{u,j} = 0 \\ 0 & \text{else} \end{cases} \qquad (6.15)$$

6.4.3   Transition Probabilities Between Non-active States

Furthermore, the probability of a transition from one non-active state u to another non-active state v, denoted by $p_N(u, v)$, is needed. It equals the probability that, of all possible transitions from state u, the one leading to state v is carried out. This means that the station inducing the transition to state v finishes its current active phase first. The probability that this station finishes first equals the probability that the processing time of that station is shorter than the processing times of all other active stations. According to Property 5.4, this can be expressed as the ratio of the transition rate between u and v and the sum of the rates of all active phases in state u.[9] The rate of the active phase inducing the transition from state u to state v equals the corresponding entry in the transition rate matrix, $Q(u, v)$. The sum of all rates of active phases in state u is given by $\lambda_{\mathcal{N}_i}(u)$. Therefore, the transition probability between the non-active states u and v, $p_N(u, v)$, results in Eq. (6.16).

$$p_N(u, v) = \frac{Q(u, v)}{\lambda_{\mathcal{N}_i}(u)} \qquad \forall u, v \in \mathcal{N}_i \qquad (6.16)$$

Note that if there is no transition from state u to state v, the entry in the transition rate matrix equals zero, $Q(u, v) = 0$. Hence, $p_N(u, v) = 0$ if there is no transition from u to v.

6.5   Determination of the Distribution

The $TBPS_i$-distribution is set up by the parameters presented in the previous section. Its transition rate matrix forms a general phase-type distribution. General phase-type distributions are introduced in Sect. 6.5.1. The assignment of the values in the transition rate matrix for the $TBPS_i$-distribution is presented in Sect. 6.5.2.

---

[9]See Property 5.4 on page 95.

6.5.1   General Phase-Type Distribution

Neuts (1982) proposed a solution method for general phase-type distributions called matrix-geometric approach. In a general phase-type distribution, transitions from any phase to any other phase are permitted rather than only the exit of the system. A graphical representation of a general phase-type distribution with n phases is depicted in Fig. 6.8.

The rate of phase i is indicated by μ_i $\forall i = 1, \ldots, n$. The probability of transiting from phase k to phase j is denoted by $p_{k,j}$ and the exit probability is labeled by $b_k = 1 - \sum_{j=1}^{n} p_{kj}$. The notation is summarized in Table 6.4. The transition rate matrix of the general phase-type distribution is given in Eq. (6.17).

$$Q^G = \begin{pmatrix} -\mu_1 & \mu_1 \cdot p_{12} & \cdots & \mu_1 \cdot p_{1n} & \mu_1 \cdot b_1 \\ \mu_2 \cdot p_{21} & -\mu_2 & \cdots & \mu_2 \cdot p_{2n} & \mu_2 \cdot b_2 \\ \vdots & & \ddots & \vdots & \vdots \\ \mu_n \cdot p_{n1} & \mu_n \cdot p_{n2} & \cdots & -\mu_n & \mu_n \cdot b_n \\ \hline 0 & 0 & \cdots & 0 & 0 \end{pmatrix} \tag{6.17}$$

The diagonal elements of Q^G contain the transition rates out of the corresponding state. All other elements constitute the transition rates between two states.[10] They are calculated by the rate of the origin state multiplied by the corresponding transition probability. The matrix Q^G is divided into the transition rates between all states, denoted by S, and the vector of the transition rates into the absorbing state, denoted by S^0, see Eq. (6.18).

$$Q^G = \begin{pmatrix} S & S^0 \\ 0 & 0 \end{pmatrix} \tag{6.18}$$

The vector σ states the probability of starting in state i, see Eq. (6.19). It indicates the so-called initial distribution.

$$\sigma = (\sigma_1, \ldots, \sigma_n) \tag{6.19}$$

The moment generating function of this general phase-type distribution is displayed in Eq. (6.20). The expected value of the random variable G is given in Eq. (6.21). The second central moment is shown in Eq. (6.22) and the variance is displayed in Eq. (6.23). These follow from the moment generating function.[11]

---

[10]For the theoretical background on the transition rate matrix, see Sect. 5.5.1.

[11]See Stewart (2009, p. 168ff).

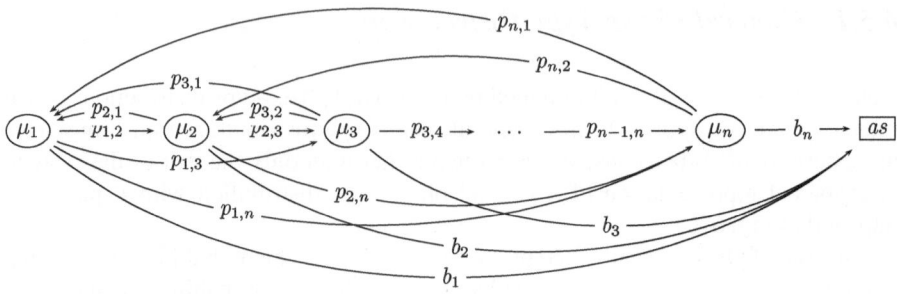

Fig. 6.8 General phase-type distribution

Table 6.4 Notation

$p_{k,j}$	Probability of transiting from phase k to phase j
b_k	Probability of exiting the phase-type distribution after phase k
S	Matrix of transition rates between states; part of matrix of Q
S^0	Vector of the transition rates into the absorbing state; part of matrix of Q
σ	Vector of the initial distribution
Q^G	Transition rate matrix of a general phase-type distribution

$$E[G^j] = (-1)^j j! \sigma S^{-j} e, \qquad j = 1, 2, \dots \qquad (6.20)$$

$$E[G] = -\sigma S^{-1} e \qquad (6.21)$$

$$E[G^2] = 2 \cdot \sigma S^{-2} e \qquad (6.22)$$

$$Var[G] = E[G^2] - E[G]^2 \qquad (6.23)$$

The probability distribution and the density function are given in Eq. (6.24) and (6.25).

$$F_G(t) = 1 - \sigma \exp^{St} e \qquad\qquad t \geq 0 \qquad (6.24)$$

$$f_G(t) = F_G'(t) = \sigma \exp^{St} S^0 \qquad\qquad t \geq 0 \qquad (6.25)$$

6.5.2 Transition Rate Matrix of the TBPS$_i$-Distribution

The *TBPS$_i$*-distribution represents a general phase-type distribution as described above. It is specified by the transition rate matrix Q^G. In the following, the computation of its elements is presented. The notation of this section is provided in Table 6.5.

Table 6.5 Notation

$p_N(u, v)$	Probability of transiting from state u to state v, where u and v are elements of the set of non-active states $\mathcal{N}_i$
$p_i^{TBPS}(d, e)$	Probability of transiting from state d into state e, where d and e are states of the $TBPS_i$-distribution
$afn(ph, u)$	Active state that leads into non-active state u when finishing the workpiece after phase ph, $afn(ph, u) \in \mathcal{A}_i^{FN}$
$tr_i^{TBPS}(d, e)$	Transition rate from state d into state e, where d and e are states of the $TBPS_i$-distribution
$\mathcal{P}_i$	Set of processing phases at station i
$\mathcal{P}_i^E$	Set of processing phases at station i with exit possibility
$\mathcal{P}_i^N$	Set of processing phases at station i without exit possibility
ph_i^{NR}	Number of phases of the phase-type distribution describing the processing time at station i
Q_i^{TBPS}	Transition rate matrix of the $TBPS_i$-distribution regarding station i

The states of the $TBPS_i$-distribution consist of the ph_i^{NR} phases from the processing time distribution at station i, which are collected in the set $\mathcal{P}_i$, and the non-active states regarding station i, $\mathcal{N}_i$. The processing phases constitute the first states of the matrix. The subsequent states represent the non-active states $\mathcal{N}_i$. Thus, the dimension of the matrix corresponds to $(|\mathcal{P}_i| + |\mathcal{N}_i|) \times (|\mathcal{P}_i| + |\mathcal{N}_i|)$. The initial distribution of the $TBPS_i$-distribution equals $\sigma_1 = 1$ and $\sigma_j = 0 \; \forall j > 1$. This allows for the fact that the time between processing-starts always begins with the first state of the processing time distribution.

The elements of the transition rate matrix of the $TBPS_i$-distribution Q_i^{TBPS} are composed of the matrix S_i and the exit vector S_i^0. In the matrix S_i, all transitions except the transitions into the absorbing state are considered. It consists of the elements $S_i(d, e)$ for all $d, e \in \mathcal{P}_i$, $\mathcal{N}_i$. The entry $S_i(d, e)$ indicates the transition rate from state d to state e. The assignment of the values for $S_i(d, e)$ is provided in Eq. (6.26).

$$S_i(d, e) = \begin{cases} -\mu_{i,d} & \text{if } d \in \mathcal{P}_i \text{ and } d = e \\ -\lambda_{\mathcal{N}_i}(d) & \text{if } d \in \mathcal{N}_i \text{ and } d = e \\ \mu_{i,d} \cdot p_i^{TBPS}(d, e) & \text{if } d \in \mathcal{P}_i \text{ and } d \neq e \\ \lambda_{\mathcal{N}_i}(d) \cdot p_i^{TBPS}(d, e) & \text{if } d \in \mathcal{N}_i \text{ and } d \neq e \end{cases} \qquad (6.26)$$

The transition rate from d to e depends on which set d and e belong to and if they are equal or not. The diagonal elements $(d = e)$ are of negative sign and constitute the departure from state d. If state d belongs to the processing time distribution, $d \in \mathcal{P}_i$, the departure rate corresponds to the rate of the d-th processing phase, $\mu_{i,d}$. If d is an element of the non-active time, $d \in \mathcal{N}_i$, the departure rate equals the rate of non-active state d, $\lambda_{\mathcal{N}_i}(d)$, see Eq. (6.14).

The transition rate from a processing phase $d \in \mathcal{P}_i$ to any other state e, with $e \neq d$, is given by the processing rate of the d-th phase, $\mu_{i,d}$, multiplied by the probability of a transition from d to e, denoted by $p_i^{TBPS}(d, e)$

(see Eq. (6.27) below). The transition rate from a non-active state d, with $d \in \mathcal{N}_i$, to any other state $e \neq d$ (more precisely, e also constitutes a non-active state), equals the rate of non-active state d, $\lambda_{\mathcal{N}_i}(d)$, again multiplied by the probability of a transition from d to e, $p_i^{TBPS}(d, e)$.

The probability of a transition from d to e, $p_i^{TBPS}(d, e)$ depends on the membership of the states to the subsets. It is given in Eq. (6.27).

$$
p^{TBPS}(d, e) = \begin{cases} (1 - a_{i,d}) \cdot \pi_{\mathcal{N}T_i}(ph_{d,i}, e) & \text{if } d \in \mathcal{P}_i^E \text{ and } e \in \mathcal{N}_i^E \\ a_{i,d} & \text{if } d, e \in \mathcal{P}_i \\ p_N(d, e) & \text{if } d, e \in \mathcal{N}_i \\ 0 & \text{else} \end{cases} \tag{6.27}
$$

If d represents a phase of the processing time which has an exit after phase d, $d \in \mathcal{P}_i^E$, and if e belongs to the non-active entrance states, $e \in \mathcal{N}_i^E$, station i transits from the active phase d into the non-active state e. This occurs with the probability $\pi_{\mathcal{N}T_i}(ph_{d,i}, e)$, given in Eq. (6.13), multiplied by the probability that the process does not continue with the next phase after phase d, $(1 - a_{i,d})$.

If both states d and e belong to the processing time, $d, e \in \mathcal{P}_i$, both d and e represent processing phases. The transition probability equals $a_{i,d}$. According to the considered phase-type distributions, it holds that $a_{i,d} = 1$ if phase d constitutes the last phase, $d = ph_i^{NR}$, or if the transition to the processing phase e is the only possibility to depart from phase d, as for example in the exponential or Erlang-k distribution. It holds that $a_{i,d} < 1$ if there is a branching after phase d, as in any Coxian distribution. If d and e are both non-active states, the transition probability equals $p_N(d, e)$ given in Eq. (6.16).

In all other cases, the transition probability equals zero. This is in particular the case if there is no exit possibility in state d, $d \in \mathcal{P}_i^N$ and $e \in \mathcal{N}_i$, even if d represents a processing state and if further e constitutes a non-active state. Another case in which the transition probability equals zero, $p_i^{TBPS}(d, e) = 0$, is prevalent, if again d constitutes a processing state and e represents a non-active state, but e does not have an entrance possibility from an active state, which means $d \in \mathcal{P}_i^E$ and $e \in \mathcal{N}_i^N$. Moreover, if the structure of the processing time distribution is such that there is no transition to a preceding phase, as in all traditional phase-type distributions, the transition probability to lower-indexed phases equals zero as well, that is for $d, e \in \mathcal{P}_i$ and $e < d$.

The vector of transitions into the absorbing state S_i^0 equals the rate of the origin state multiplied by the probability of transiting into the absorbing state. That probability corresponds to the counter-probability of transiting into any state of the $TBPS_i$-distribution. It is given in Eq. (6.28).

$$
S_i^0(d) = \begin{cases} \mu_{i,d} \cdot (1 - a_{i,d}) \cdot (1 - \pi_{\mathcal{N}_i}) & \text{if } d \in \mathcal{P}_i \\ \lambda_{\mathcal{N}_i}(d) \cdot (1 - \sum_{u=1}^n p_N(d, u)) & \text{if } d \in \mathcal{N}_i \end{cases} \tag{6.28}
$$

If the origin state d constitutes a processing state, $d \in \mathcal{P}_i$, the rate with which $TBPS_i$ is finished equals the processing rate of phase d, $\mu_{i,d}$, multiplied by the probability of not continuing with the next phase $(1 - a_{i,d})$ and the probability of not transiting into non-activity, $(1 - \pi_{\mathcal{N}_i})$. If d represents a non-active state, the rate of exiting $TBPS_i$ equals the rate of that state, $\lambda_{\mathcal{N}_i}(d)$, multiplied by the probability of not transiting into any other non-active state, $(1 - \sum_{u=1}^{n} p_N(d,u))$.

To provide an example, the transition rate matrix Q_i^{TBPS} with a Cox-2 distribution as processing time, as depicted in Fig. 6.7, is expressed by S_i and S_i^0 in Eqs. (6.29) and (6.30), respectively.

$$
S_i = \begin{array}{cccccc}
{\scriptstyle ph=1} & {\scriptstyle ph=2} & {\scriptstyle u=1} & {\scriptstyle u=2} & & {\scriptstyle u=n} \\
\left(\begin{array}{cccccc}
-\mu_{i1} & \mu_{i1} \cdot a_i & \mu_{i1} \cdot (1-a_i) \cdot \pi_{\mathcal{N}\mathcal{T}_i}(1) & \mu_{i1}(1-a_i) \cdot \pi_{\mathcal{N}\mathcal{T}_i}(2) & \cdots & \mu_{i1} \cdot (1-a_i) \cdot \pi_{\mathcal{N}\mathcal{T}_i}(n) \\
0 & -\mu_{i2} & \mu_{i2} \cdot \pi_{\mathcal{N}\mathcal{T}_i}(1) & \mu_{i2} \cdot \pi_{\mathcal{N}\mathcal{T}_i}(2) & \cdots & \mu_{i2} \cdot \pi_{\mathcal{N}\mathcal{T}_i}(n) \\
0 & 0 & -\lambda_{\mathcal{N}_i}(1) & \lambda_{\mathcal{N}_i}(1) \cdot p_{1,2} & \cdots & \lambda_{\mathcal{N}_i}(1) \cdot p_{1,n} \\
0 & 0 & \lambda_{\mathcal{N}_i}(2) \cdot p_{2,1} & -\lambda_{\mathcal{N}_i}(2) & \cdots & \lambda_{\mathcal{N}_i}(2) \cdot p_{2,n} \\
\vdots & \vdots & \vdots & & \ddots & \vdots \\
0 & 0 & \lambda_{\mathcal{N}_i}(n) \cdot p_{n,2} & \cdots & & -\lambda_{\mathcal{N}_i}(n)
\end{array}\right)
\end{array}
$$

$$(6.29)$$

$$
S_i^0 = \begin{pmatrix}
\mu_{i1} \cdot (1-a_i) \cdot (1-\pi_{\mathcal{N}_i}) \\
\mu_{i2} \cdot (1-\pi_{\mathcal{N}_i}) \\
\lambda_{\mathcal{N}_i}(1) \cdot (1 - \sum_{j=1}^{n} p_{1,j}) \\
\lambda_{\mathcal{N}_i}(2) \cdot (1 - \sum_{j=1}^{n} p_{2,j}) \\
\vdots \\
\lambda_{\mathcal{N}_i}(n) \cdot (1 - \sum_{j=1}^{n} p_{n,j})
\end{pmatrix}
$$

$$(6.30)$$

6.6 Examples

In this section, two examples are provided. In the first example, a CQN with exponentially distributed processing times is considered, see Sect. 6.6.1. The other example illustrates the modeling of the $TBPS_i$-distribution of a CQN with Cox-2 and hypo-exponential-2 distributed processing times, see Sect. 6.6.2.

6.6.1 Exponential Distribution

In this example, we consider a three-station closed queueing network with $n = 3$ customers. The service times are exponentially distributed with the rates μ_1, μ_2, and μ_3. To provide numerical results, we set $\mu_1 = 0.5$, $\mu_2 = 0.7$, and $\mu_3 = 0.9$. The buffer in front of each station contains one unit of buffer space, $b_i = 1 \ \forall i$. For the calculation of the $TBPS_i$-distribution, we consider station $i = 1$. Figure 6.9 depicts the Markov chain of that system.

Fig. 6.9 Markov chain of a three-station system with exponential distribution and blocking

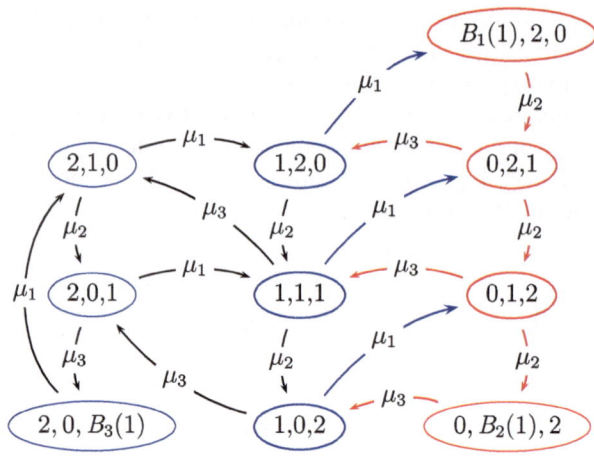

The colors of the states in Fig. 6.9 correspond to the subsets. The non-active entrance states, $\mathcal{N}_1^E$, are highlighted in thick red, whereas the non-active non-entrance states $\mathcal{N}_1^N$ are circled in thin red. The active states that finish after phase ph and exit into non-activity, $\mathcal{A}_1^{FN}$ are marked in thick blue, and those that do not transit into non-activity, $\{\mathcal{A}_1^F \setminus \mathcal{A}_1^{FN}\}$, are indicated by a thin blue circle.

The subsets are given in Eqs. (6.31)–(6.34).

$$\mathcal{A}_1^F = \{(1,2,0),(1,1,1),(1,0,2),(2,1,0),(2,0,1),(2,0,B_3(1))\} \tag{6.31}$$

$$\mathcal{A}_1^{FN} = \{(1,2,0),(1,1,1),(1,0,2)\} \tag{6.32}$$

$$\mathcal{N}_1^E = \{(B_1(1),2,0),(0,2,1),(0,1,2)\} \tag{6.33}$$

$$\mathcal{N}_1^N = \{(0,B_2(1),2)\} \tag{6.34}$$

The probability of transiting into non-activity at station 1 corresponds to Eq. (6.10) and is given in Eq. (6.35).

$$
\pi_{\mathcal{N}_1} = \frac{\displaystyle\sum_{s \in \mathcal{A}_i^{FN}} \pi(s)}{\displaystyle\sum_{t \in \mathcal{A}_i^F} \pi(t)}
$$

$$
= \frac{\pi(1,2,0) + \pi(1,1,1) + \pi(1,0,2)}{\pi(1,2,0) + \pi(1,1,1) + \pi(1,0,2) + \pi(2,1,0) + \pi(2,0,1) + \pi(2,0,B_3(1))}
$$

$$
= \frac{0.1578 + 0.0877 + 0.0487}{0.1578 + 0.0877 + 0.0487 + 0.1127 + 0.0626 + 0.0805} = 0.5348. \tag{6.35}
$$

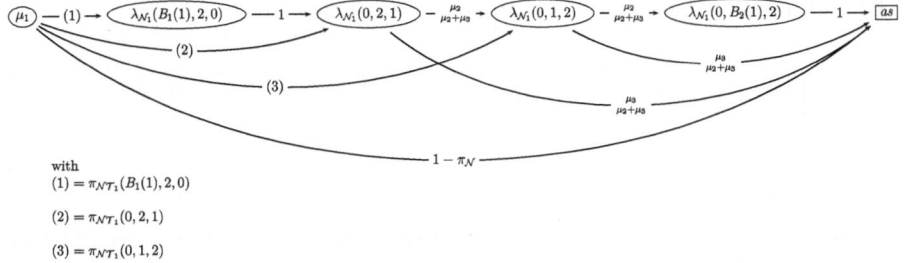

with
$(1) = \pi_{\mathcal{N}T_1}(B_1(1), 2, 0)$

$(2) = \pi_{\mathcal{N}T_1}(0, 2, 1)$

$(3) = \pi_{\mathcal{N}T_1}(0, 1, 2)$

Fig. 6.10 $TBPS_i$-distribution at station 1 for the example of Fig. 6.9

The probabilities of entering a particular non-active state conform with Eq. (6.12) and are given in Eqs. (6.36)–(6.38).

$$\pi_{\mathcal{N}T_1}(1, (B_1(1), 2, 0)) = \frac{\pi(1, 2, 0)}{\sum_{t \in \mathcal{A}_i^F} \pi(t)} \qquad = 0.2869 \qquad (6.36)$$

$$\pi_{\mathcal{N}T_1}(1, (0, 2, 1)) = \frac{\pi(1, 1, 1)}{\sum_{t \in \mathcal{A}_i^F} \pi(t)} \qquad = 0.15939 \qquad (6.37)$$

$$\pi_{\mathcal{N}T_1}(1, (0, 1, 2)) = \frac{\pi(1, 0, 2)}{\sum_{t \in \mathcal{A}_i^F} \pi(t)} \qquad = 0.08855 \qquad (6.38)$$

As it is supposed to be,

$$\sum_{u \in \mathcal{N}_1^E} \pi_{\mathcal{N}T_1}(1, u) = 0.2869 + 0.15939 + 0.08855 = 0.5348 = \pi_{\mathcal{N}_1}. \qquad (6.39)$$

Figure 6.10 displays the $TBPS_i$-distribution of the example. The exponential rates of the non-active phases are given in Eqs. (6.40)–(6.43).

$$\lambda_{\mathcal{N}_1}(B_1(1), 2, 0) = \mu_2 = 0.5 \qquad (6.40)$$

$$\lambda_{\mathcal{N}_1}(0, 2, 1) = \mu_2 + \mu_3 = 1.4 \qquad (6.41)$$

$$\lambda_{\mathcal{N}_1}(0, 1, 2) = \mu_2 + \mu_3 = 1.4 \qquad (6.42)$$

$$\lambda_{\mathcal{N}_1}(0, B_2(1), 2) = \mu_3 = 0.9 \qquad (6.43)$$

The transition rate matrix of this example is given below.

$$
Q_i^{TBPS} = \begin{pmatrix}
-\mu_1 & \mu_1 \cdot \pi_{\mathcal{NT}_1}(B_1(1),2,0) & \mu_1 \cdot \pi_{\mathcal{NT}_1}(0,2,1) & \mu_1 \cdot \pi_{\mathcal{NT}_1}(0,1,2) & 0 & \mu_1 \cdot (1 - \pi_{\mathcal{N}_1}) \\
0 & -\mu_2 & \mu_2 & 0 & 0 & 0 \\
0 & 0 & -(\mu_2 + \mu_3) & \mu_2 & 0 & \mu_3 \\
0 & 0 & 0 & -(\mu_2 + \mu_3) & \mu_2 & \mu_3 \\
0 & 0 & 0 & 0 & -\mu_3 & \mu_3 \\
0 & 0 & 0 & 0 & 0 & 0
\end{pmatrix}
$$

The mean, the variance, and the coefficient of variation of $TBPS_i$ are obtained from Eqs. (6.20) to (6.23) and are displayed in Eqs. (6.44)–(6.46).

$$E[TBPS_i] = 2.5966 \tag{6.44}$$

$$\text{Var}[TBPS_i] = 5.5674 \tag{6.45}$$

$$\text{CV}[TBPS_i] = 0.9087 \tag{6.46}$$

6.6.2 Cox-2 Distribution

The second example serves to point out how $TBPS_i$ is set up for a CQN with processing times of more than one phase and exit. Therefore, a three-station closed queueing system with a Cox-2 distribution at station 1 and a hypo-exponential-2 distribution at stations 2 and 3 is considered. Each station has a buffer capacity of one unit, $b_i = 1 \, \forall i. \, n = 2$ customers circle in the system. In Fig. 6.11, the Markov chain of this example is depicted.

The colors of the states in Fig. 6.11 correspond to the subsets in the same way as in the previous example. The non-active entrance states, $\mathcal{N}_i^E$, are highlighted in thick red, whereas the non-active non-entrance states $\mathcal{N}_i^N$ are circled in thin red. The active states that finish after phase $ph = 1$ and exit into non-activity, $\mathcal{A}_i^{FN}(1)$ are marked in thick blue, and those that do not transit into non-activity, $\{\mathcal{A}_i^F(1) \setminus \mathcal{A}_i^{FN}(1)\}$, are indicated by a thin blue circle.

The subsets of state-classifications regarding station $i = 1$ and phase $ph = 1$ are given in Eqs. (6.47)–(6.50).

$$
\begin{aligned}
\mathcal{A}_i^F(1) =\{&([0,1],[0,0],[0,1]), ([0,1],[0,0],[0,2]), \\
&([1,1],[0,0],[0,0]), ([0,1],[0,1],[0,0]), \\
&([0,1],[0,2],[0,0])\}
\end{aligned} \tag{6.47}
$$

$$
\begin{aligned}
\mathcal{A}_i^{FN}(1) =\{&([0,1],[0,0],[0,1]), ([0,1],[0,0],[0,2]), \\
&([0,1],[0,1],[0,0]), ([0,1],[0,2],[0,0])\}
\end{aligned} \tag{6.48}
$$

$$
\begin{aligned}
\mathcal{N}_i^E =\{&([0,0],[0,2],[0,2]), ([0,0],[0,1],[0,1]), \\
&([0,0],[1,1],[0,0]), ([0,0],[1,2],[0,0])\}
\end{aligned} \tag{6.49}
$$

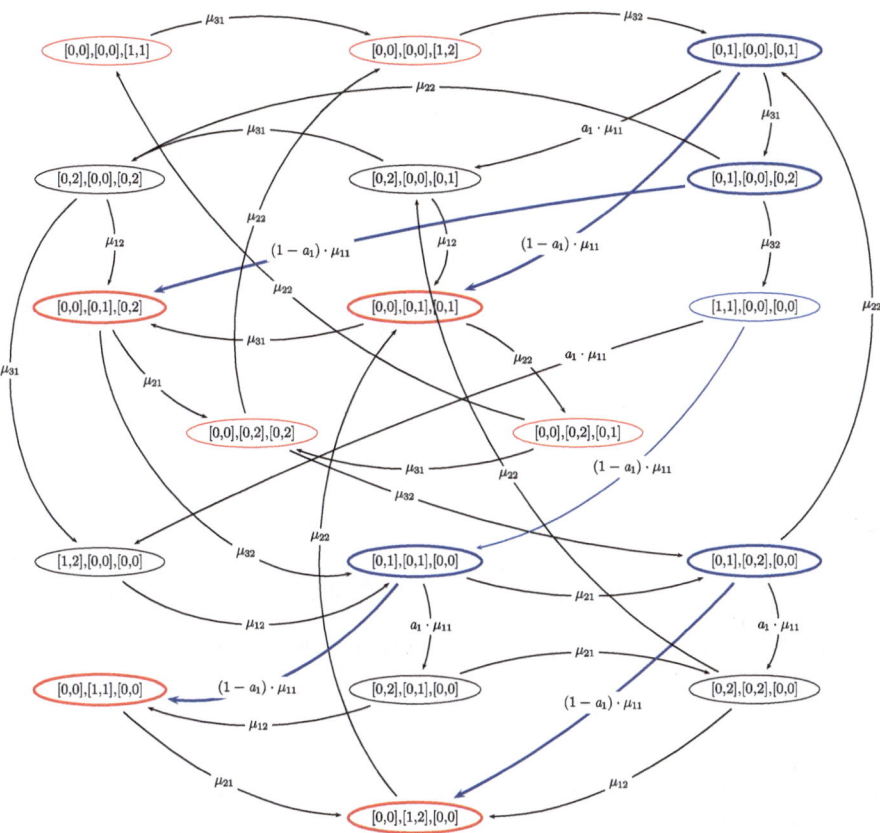

Fig. 6.11 Markov chain of a three-station system with Cox-2 and hypo-exponential-2 distribution

$$\mathcal{N}_i^N = \{([0,0],[0,0],[1,1]),\ ([0,0],[0,0],[1,2]),$$
$$([0,0],[0,2],[0,2]),\ ([0,0],[0,2],[0,1])\} \qquad (6.50)$$

The resulting $TBPS_i$-distribution is presented in Fig. 6.12.

6.7   Numerical Study of the Parameter Influences on $CV^2(TBPS_i)$

The influences of the configuration-parameters on the reciprocal of the expected value of $TBPS_i$, the production rate, have been well studied. With the approach proposed above, the analytical computation of higher-order moments is enabled.

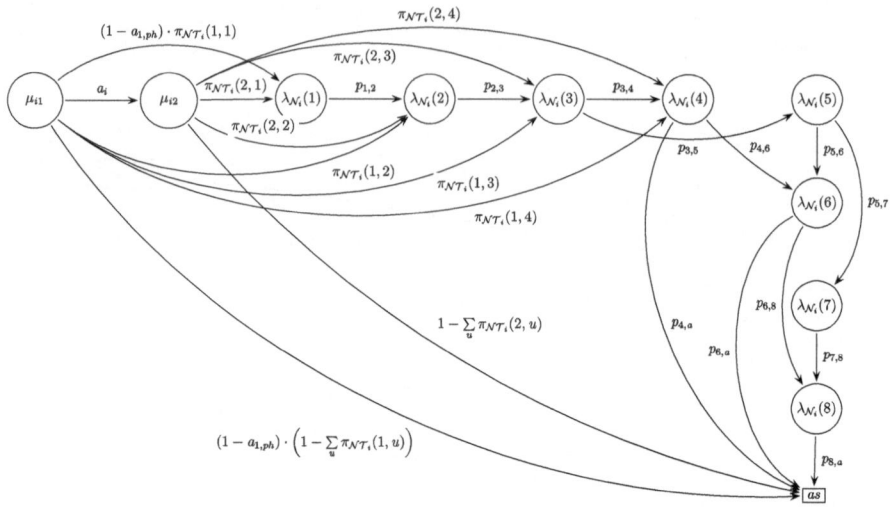

Legend of states:
(1) [0,0],[1,1],[0,0]   (2) [0,0],[1,2],[0,0]   (3) [0,0],[0,1],[0,1]   (4) [0,0],[0,1],[0,2]
(5) [0,0],[0,2],[0,1]   (6) [0,0],[0,2],[0,2]   (7) [0,0],[0,0],[1,1]   (8) [0,0],[0,0],[1,2]

Fig. 6.12 Time between processing starts at station 1 with Cox-2 and hypo-exponential-2 distributed processing times

In the following, we will investigate the influences of configuration parameters on the squared coefficient of variation of $TBPS_i$, $CV^2(TBPS_i)$, by means of a numerical study. We varied the processing rates, the coefficients of variation of the processing time, the buffer capacities, and the number of workpieces.

6.7.1   Influence of the Processing Rate

The tested values of the processing rate and its labels are provided in Table 6.6. The other parameter values were set to $b = \{1, 1, 1, 1\}$, $c^2 = \{0.64, 0.64, 0.64, 0.64\}$, and $n = 1, \ldots, N$. The results of the $CV^2(TBPS_i)$ for the above defined test set are depicted in Fig. 6.13. As can be seen, the $CV^2(TBPS_i)$ is equal for different levels of μ, as long as the ratio over all stations is the same: The $CV^2(TBPS_i)$ is equal for mu-4 and mu-5, so is the $CV^2(TBPS_i)$ for mu-2 and mu-3. The $CV^2(TBPS_i)$ of mu-4 and mu-5 is higher than the $CV^2(TBPS_i)$ of mu-2 and mu-3. That shows that unbalance of the processing rates increases the variability of the time between processing starts in CQN.

Table 6.6 Tested values of the processing rate

Values	Label
$\mu = \{0.7, 0.5, 0.9, 0.8\}$	mu-1
$\mu = \{0.5, 0.5, 0.5, 0.5\}$	mu-2
$\mu = \{3, 3, 3, 3\}$	mu-3
$\mu = \{1, 3, 1, 3\}$	mu-4
$\mu = \{2, 6, 2, 6\}$	mu-5

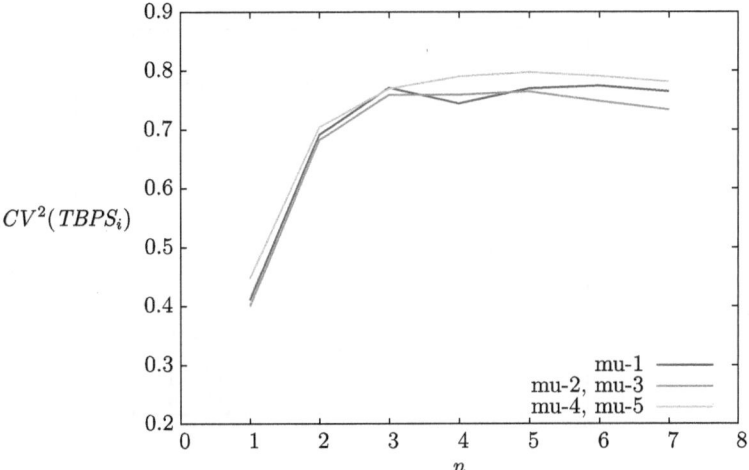

Fig. 6.13 $CV^2(TBPS_i)$ subject to n: different processing rates

6.7.2 Influence of c^2

In the following, we test the influence of both the level and the balance of the coefficient of variation of the processing time c^2 on the $CV^2(TBPS_i)$.

6.7.2.1 Influence of the Level of c^2

The tested coefficients of variation of the processing time c^2 range from low to high and are equal at all stations. The values are given in Table 6.7. The other parameter values were set to $b = \{2, 2, 2\}$, $\mu = \{0.7, 0.5, 0.9\}$, and $n = 1, \ldots, N$. The results of the production rate and the coefficient of variation of $TBPS_i$ for the above defined test set are depicted in Figs. 6.14 and 6.15. The graphics show that the higher the squared coefficient of variation of the processing time is, the higher is the coefficient of variation of the time between processing starts, $CV^2(TBPS_i)$. A higher c^2 also implies a lower PR. Further, the $CV^2(TBPS_i)$ depends on n: $CV^2(TBPS_i)$ is lower for very low and very high n.

Table 6.7 Tested values of c^2 (level)

Values
$c^2 = \{0.125, 0.125, 0.125\}$
$c^2 = \{0.25, 0.25, 0.25\}$
$c^2 = \{0.64, 0.64, 0.64\}$
$c^2 = \{1, 1, 1\}$
$c^2 = \{2, 2, 2\}$

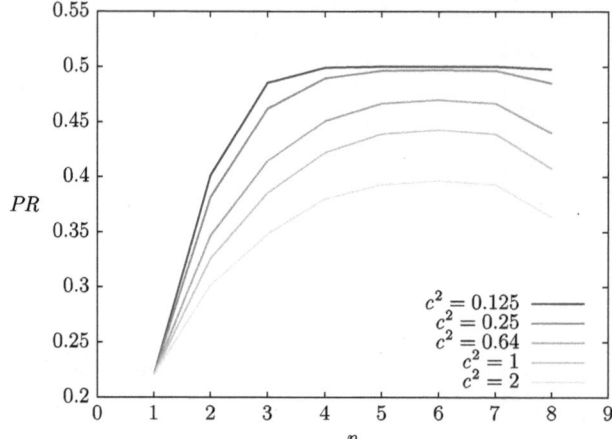

Fig. 6.14 Production rate subject to n: level of c^2

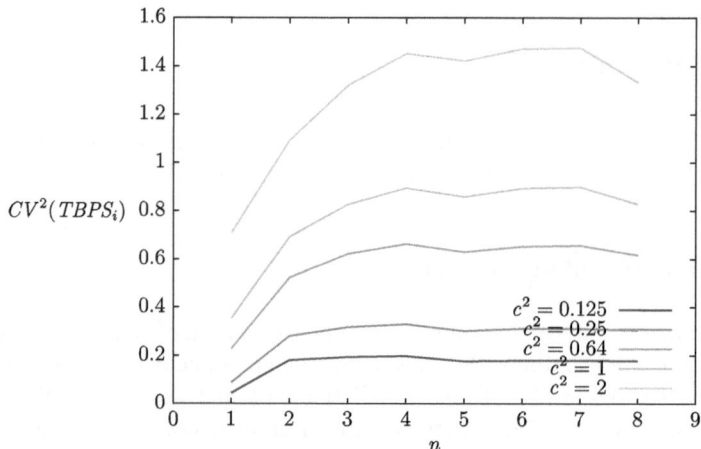

Fig. 6.15 $CV^2(TBPS_i)$ subject to n: level of c^2

Table 6.8 Tested values of c^2 (balance)

Values	Label
$c^2 = \{1, 0.64, 0.8, 1.6\}$	C-UB-l
$c^2 = \{2, 3, 2, 2\}$	C-UB-h1
$c^2 = \{2, 4, 2, 4\}$	C-UB-h2
$c^2 = \{0.64, 0.64, 0.64, 0.64\}$	C-B-l
$c^2 = \{1, 1, 1, 1\}$	C-B-m
$c^2 = \{2, 2, 2, 2\}$	C-B-h

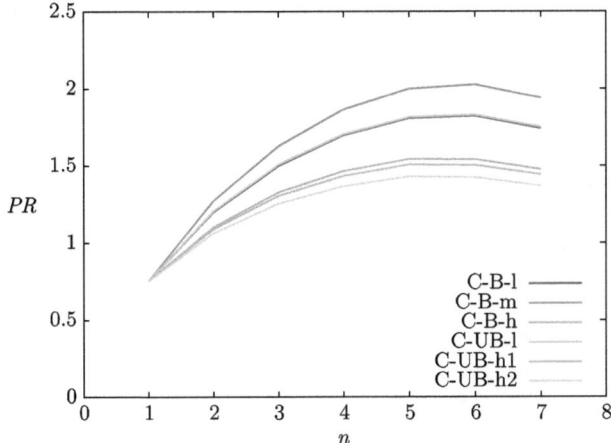

Fig. 6.16 Production rate subject to n: balance of c^2

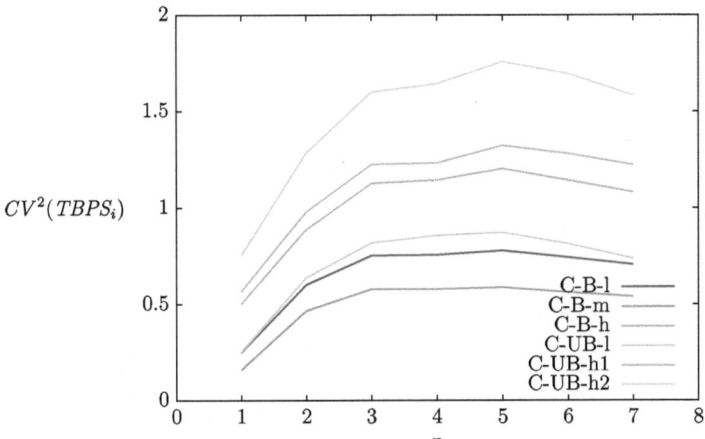

Fig. 6.17 $CV^2(TBPS_i)$ subject to n: balance of c^2

Table 6.9 Tested values of the buffer capacities

Values	Label
$b = \{1, 1, 1, 1\}$	B-b-l
$b = \{3, 3, 3, 3\}$	B-b-h
$b = \{2, 4, 3, 3\}$	B-u-l
$b = \{1, 5, 1, 5\}$	B-u-h

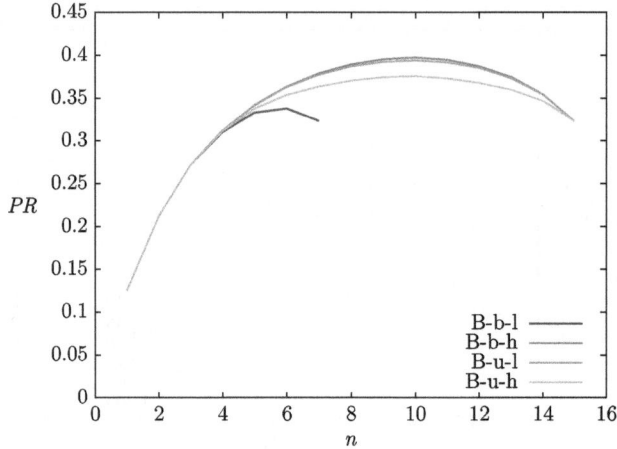

Fig. 6.18 Production rate subject to n: balance of buffer sizes

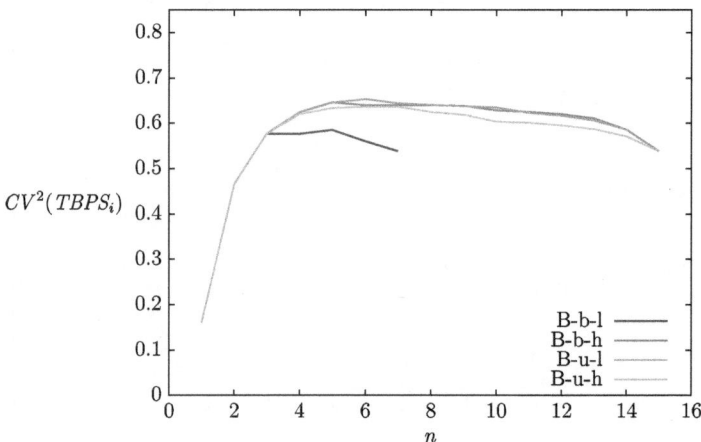

Fig. 6.19 $CV^2(TBPS_i)$ subject to n: balance of buffer sizes

If c^2 is small ($c^2 = 0.125$), $CV^2(TBPS_i) > c^2$. If c^2 is high ($c^2 = 1$ and $c^2 = 2$), $CV^2(TBPS_i) < c^2$. A very high variability of the processing time is captured in the network and, therefore, the $CV^2(TBPS_i)$ is lowered, whereas a low variability of the processing time results in a higher variability of the $TBPS_i$.

6.7.2.2 Influence of the Balance of c^2

The tested coefficients of variation of the processing time are balanced or unbalanced over the stations, see Table 6.8. The other parameter values were again set to $b = \{1, 1, 1, 1\}$, $\mu = \{3, 3, 3, 3\}$, and $n = 1, \ldots, N$.

The results of the PR and the $CV^2(TBPS_i)$ for the above defined test set are depicted in Figs. 6.16 and 6.17.

From these graphics, we find: If the coefficient of variation at one station is higher than a certain level $\bar{c}$, whereas those of the other stations are equal or lower than $\bar{c}$, $TBPS_i$ is more variable. The production rate, however, is almost identical in these two cases. This shows a comparison of the instance C-UB-l (in which one c^2 is higher than 1 (1.6) and the other c^2 are lower or equal) to C-B-m (in which all c^2 are equal to 1) and a comparison of C-UB-h1 (in which one c^2 higher than 2 (3) and all other c^2 are equal to 2) to C-B-h (in which all c^2 are equal to 2). Therefore, the more c^2 varies between the stations, the higher is $CV^2(TBPS_i)$.

6.7.3 Influence of the Buffer Capacity

The tested buffer capacities are provided in Table 6.9. The other parameter values are $\mu = \{0.5, 0.5, 0.5, 0.5\}$, $c^2 = \{0.64, 0.64, 0.64, 0.64\}$, and $n = 1, \ldots, N$. The results of the production rate and the coefficient of variation of $TBPS_i$ for the defined test set are depicted in Figs. 6.18 and 6.19. It shows that the more unbalanced the buffer capacity is, the lower is the $CV^2(TBPS_i)$ (because of a distinct bottleneck).

Chapter 7
Conclusion

We investigated closed queueing networks with general processing times and finite buffer capacities. The processing time distribution was specified by the first two moments or by phase-type distributions. Under these assumptions, an approximate and an exact approach for the performance analysis of the first-moment performance measures were proposed. Moreover, an exact method for the determination of the distribution of the time between processing starts of CQN was derived.

In the approximate approach, the CQN was modeled as open system. For this a virtual arrival rate was found so that the work-in-process of the open system corresponds to the number of workpieces in the closed queueing system. The open system was evaluated by a decomposition approach. The resulting production-rate estimate of the open system served as an approximation for that of the closed system. In the numerical study, the results of the algorithm were compared to simulation results and to results of existent procedures. As shown, the algorithm is fast and the approximation quality is high. According to the construction of the approach, the production-rate estimate depends on the work-in-process approximation. The WIP-estimate is precise, especially in systems with low coefficients of variation, large buffers, or many stations. Regarding the range of workpieces, the decomposition approach works best for the most relevant area of the number of workpieces, which is in the medium range.

In the exact approach, CQN were modeled as Markov chains. The processing times were represented by phase-type distributions. Blocking was modeled in blocking states corresponding to the phase-type distribution. The Markov chain was set up so that all interactions between stations were considered. The performance measures were obtained by setting up and solving a single Markov chain for the entire queueing system. The computation time depends—assuming the computer equipment to be given—on the size of the transition rate matrix of the Markov chain. The matrix is larger, the more stations the system contains, the more phases the distribution consists of, and the larger the buffer capacities are. As shown, the method is very fast for processing time distributions consisting of one or two phases (that means, for squared coefficients of variation greater than 0.5). The computation

S. Lagershausen, *Performance Analysis of Closed Queueing Networks*, Lecture Notes in Economics and Mathematical Systems 663, DOI 10.1007/978-3-642-32214-3_7, © Springer-Verlag Berlin Heidelberg 2013

time is also very low for two-station systems, independent of the processing time distribution.

The exact distribution of the time between processing starts is represented by a general phase-type distribution. It is modeled as processing time that proceeds—with the probability of transiting into non-activity—with non-active time. The transition probabilities and the non-active states are derived from the steady-state solution of the Markov-chain analysis. The transition rate matrix that specifies the general phase-type distribution is set up according to the transition probabilities and the transition rates of both the processing and the non-active states. In a numerical study, the influence of the configuration parameters on the coefficient of variation of the time between processing starts was studied. The CV is lower, the more balanced the processing rates and the lower the CVs of the processing times are. Further, the CV is lower for a small or a high number of workpieces.

The proposed approaches may be extended to multiple servers, machine failures, and to other service disciplines, blocking mechanisms, or topologies. The decomposition approach may be expanded to consider closed assembly and disassembly systems using appropriate queueing systems. The Markov chain approach could, for example, be applied to open networks. The concept for the distribution of the time between processing starts can be used to derive the inter-departure-time distribution. The modeling of open two-station subsystems by Markov chains together with the calculation of the inter-departure-time distribution would enable the exact analysis of subsystems. This could be used as a building block in decomposition approaches.

Bibliography

Agrawal, S., Buzen, J., & Shum, A. (1984). Response time preservation – A general technique for developing approximate algorithms for queueing networks. *Association for Computing Machinery Sigmetrics Performance Evaluation Review, 12*(3), 63–77.

Akyildiz, I. F. (1985). Die erweiterte parametrische Analyse für geschlossene Warteschlangennetze. In *Messung, Modellierung und Bewertung von Rechensystemen, 3. GI/NTG-Fachtagung* (pp. 170–185), Dortmund.

Akyildiz, I. F. (1987). Exact product form solution for queueing networks with blocking. *IEEE Transactions on Computers, 36*(1), 122–125.

Akyildiz, I. F. (1988a). General closed queueing networks with blocking. In *Proceedings of the 12th IFIP WG 7.3 International Symposium on Computer Performance Modelling, Measurement and Evaluation* (pp. 283–303), Brussels.

Akyildiz, I. F. (1988b). Mean value analysis for blocking queueing networks. *IEEE Transaction on Software Engineering, 14*(4), 418–428.

Akyildiz, I. F. (1988c). On the exact and approximate throughput analysis of closed queueing networks with blocking. *IEEE Transaction on Software Engineering, 14*(1), 62–70.

Akyildiz, I. F. (1989). Product form approximations for queueing networks with multiple servers and blocking. *IEEE Transactions on Computers, 38*(1), 99–114.

Akyildiz, I. F., & Bolch, G. (1988). Mean value analysis approximation for multiple server queueing networks. *Performance Evaluation, 8*(2), 77–91.

Akyildiz, I. F., & Huang, C.-C. (1993). Exact analysis of queueing networks with multiple job classes and blocking-after-service. *Queueing Systems, 13*(4), 427–440.

Akyildiz, I. F., & Liebeherr, J. (1989). Application of norton's theorem on queueing networks with finite capacities. In *Infocom '89. Proceedings of the Eighth Annual Joint Conference of the IEEE Computer and Communications Societies* (Vol. 3, pp. 914–923). Washington, D.C.: IEEE Computer Society Press.

Akyildiz, I. F., & Sieber, A. (1988). Approximate analysis of load dependent general queueing networks. *IEEE Transaction on Software Engineering, 14*(11), 1537–1545.

Akyildiz, I. F., & von Brand, H. (1994). Exact solutions for networks of queues with blocking-after-service. *Theoretical Computer Science, 125*(1), 111–130.

Allen, A. O. (1990). *Probability, statistics, and queueing theory*. Boston: Academic.

Altiok, T. (1985). On the phase-type approximations of general distributions. *IIE Transactions, 17*(2), 110–116.

Altiok, T. (1996). *Performance analysis of manufacturing systems*. New York: Springer.

Altiok, T., & Ranjan, R. (1989). Analysis of production lines with genral service times and finite buffers: A two-node decomposition approach. *Engineering Cost and Production Economics, 17*(1–4), 155–165.

S. Lagershausen, *Performance Analysis of Closed Queueing Networks*, Lecture Notes in Economics and Mathematical Systems 663, DOI 10.1007/978-3-642-32214-3, © Springer-Verlag Berlin Heidelberg 2013

Asmussen, S., Nerman, O., & Olsson, M. (1996). Fitting phase-type distributions via the EM algorithm. *Scandinavian Journal of Statistics, 23*(4), 419–441.

Balsamo, S., & Clò, M. (1998). A convolution algorithm for product-form queueing networks with blocking. *Annals of Operations Research, 79*(0), 97–117.

Balsamo, S., & Donatiello, L. (1989). On the cycle time distribution in a two-stage cyclic network with blocking. *IEEE Transactions on Software Engineering, 15*(10), 1206–1216.

Balsamo, S., Personè, V. D. N., & Inverardi, P. (2003). A review on queueing network models with finite capacity queues for software architectures performance prediction. *Performance Evaluation, 51*(2–4), 269–288.

Bard, Y. (1979). Some extensions to multiclass queueing network analysis. In *Proceedings of 4th International Symposium on Modelling and Performance Evaluation of Computer Systems* (pp. 51–62), Vienna.

Bard, Y. (1981). A simple approach to system modeling. *Performance Evaluation, 1*(3), 225–248.

Baskett, F., Chandy, K. M., Muntz, R. R., & Palacios, F. G. (1975). Open, closed, and mixed networks of queues with different classes of customers. *Journal of the Association for Computing Machinery, 22*(2), 248–260.

Baynat, B., & Dallery, Y. (1993). Approximate techniques for general closed queueing networks with subnetworks having population constraints. *European Journal of Operational Research, 69*(2), 250–264.

Baynat, B., & Dallery, Y. (1996). A product-form approximation method for general closed queueing networks with several classes of customers. *Performance Evaluation, 24*(3), 165–188.

Baynat, B., Dallery, Y., & Ross, K. (1994). A decomposition approximation method for multiclass bcmp queueing networks with multiple-server stations. *Annals of Operations Research, 48*(3), 273–294.

Bolch, G., & Fischer, M. (1993). Bottapprox: Eine Engpaßanalyse für geschlossene Warteschlangennetze auf der Basis der Summationsmethode. In *Proceedings of 4th GI/ITG Conference on Measurement, Modeling and Performance Evaluation of Computer and Communication Systems* (pp. 511–517).

Bolch, G., Fleischmann, G., & Schreppel, R. (1987). Ein funktionales Konzept zur Analyse von Warteschlangennetzen und Optmierung von Leistungsgrößen. In *Proceedings of 4th GI/ITG Conference on Measurement, Modeling and Performance Evaluation of Computer and Communication Systems* (pp. 327–342).

Bolch, G., Greiner, S., de Meer, H., & Trivedi, K. S. (2006). *Queueing networks and Markov chains. Modeling and performance evaluation with computer science applications* (2nd ed.). Hoboken: Wiley.

Bouhchouch, A., Frein, Y., & Dallery, Y. (1993). Analysis of closed-loop manufacturing systems with finite buffers. *Applied Stochastic Models and Data Analysis, 9*(2), 111–125.

Bouhchouch, A., Frein, Y., & Dallery, Y. (1996). Performance evaluation of closed tandem queueing networks with finite buffers. *Performance Evaluation, 26*(2), 115–132.

Boxma, O. J. (1983). The cyclic queue with one general and one exponential server. *Advances in Applied Probability, 15*(4), 857–873.

Boxma, O. J., & Donk, P. (1982). On response time and cycle time distribution in a two-stage cyclic queue. *Performance Evaluation, 2*(3), 181–194.

Boxma, O. J., Kelly, F. P., & Konheim, A. G. (1984). The product form for sojourn time distributions in cyclic exponential queues. *Journal of the Assiciation for Computing Machinery, 31*(1), 128–133.

Buzacott, J. A., Liu, X. G., & Shanthikumar, J. G. (1995). Multistage flow line analysis with the stopped arrival queue model. *IIE Transactions, 27*(4), 444–455.

Buzacott, J. A., & Shanthikumar, J. G. (1993). *Stochastic models of manufacturing systems*. Englewood Cliffs: Prentice Hall.

Buzacott, J. A., & Yao, D. D. (1986). Flexible manufacturinig systems: A review of analytical models. *Management Science, 32*(7), 890–905.

Buzen, J. (1973). Computational algorithms for closed queueing systems with exponential servers. *Communications of the Association for Computing Machinery, 16*(9), 527–531.

Carbini, S., Donatiello, L., & Iazeolla, G. (1986). An efficient algorithm for the cycle time distribution in two-stage cyclic queues with a non-exponential server. In *Proceedings of the International Seminar in Teletraffic Analysis and Computer Performance Evaluation* (pp. 99–115), Amsterdam.

Cavaille, J.-B., & Dubois, D. (1982). Heuristic methods based on mean value analysis for flexible manufacturing system performance evaluation. In *Proceedings of the 21st IEEE Conference on Decision and Control* (pp. 1061–1066), Orlando.

Chandy, K. M., Herzog, U., & Woo, L. (1975a). Approximate analysis of general queuing networks. *IBM Journal of Research and Development, 19*(1), 43–49.

Chandy, K. M., Herzog, U., & Woo, L. (1975b). Parametric analysis of queuing networks. *IBM Journal of Research and Development, 19*(1), 36–42.

Chandy, K. M., & Neuse, D. (1982). Linearizer: A heuristic algorithm for queueing network models of computing systems. *Communications of The Association for Computing Machinery, 25*(2), 126–134.

Chandy, K. M., & Sauer, C. (1980). Computational algorithms for product form queueing networks. *Communications of the Association for Computing Machinery, 23*(10), 573–583.

Chen, C., George, E., & Tardif, V. (2001). A bayesian model of cycle time prediction. *IIE Transactions, 33*(10), 921–930.

Chen-Hong, C. (1999). *Cycle time modeling*. Ph.D. thesis, University of Texas, Austin.

Chow, W.-M. (1980). The cycle time distribution of exponential cyclic queues. *Journal of the Association for Computing Machinery, 27*(2), 281–286.

Clò, M. (1998). MVA for product-form cyclic queueing networks with blocking. *Annals of Operations Research, 79*(0), 83–96.

Cox, D. R. (1955). A use of complex probabilities in the theory of stochastic processes. *Mathematical Proceedings of the Cambridge Philosophical Society, 51*(2), 313–319.

Curry, G. L., & Feldman, R. M. (2008). *Manufacturing systems modeling and analysis*. Berlin: Springer.

Daduna, H. (1984). Burkes's theorem in passage times in gordon-newell networks. *Advances in Applied Probability, 16*(4), 867–886.

Daduna, H. (1986). Two-stage cyclic queues with nonexponential servers: Steady-state and cyclic time. *Operations Research, 34*(3), 455–459.

Dallery, Y., & Cao, X.-R. (1992). Operational analysis of stochastic closed queueing networks. *Performance Evaluation, 14*(1), 43–61.

Dallery, Y., & Frein, Y. (1989). A decomposition method for the approximate analysis of closed queueing networks with blocking. In *Proceedings of the First International Workshop on Queueing Networks with Blocking* (pp. 193–209). North Carolina.

Dallery, Y., & Gershwin, S. B. (1992). Manufacturing flow line systems: A review of models and analytical results. *Queueing Systems, 12*(1–2), 3–94.

Dayar, T., & Meri, A. (2008). Kronecker representation and decompositional analysis of closed queueing networks with phase-type service distributions and arbitrary buffer sizes. *Annals of Operations Research, 164*(1), 193–210.

Disney, R. L., & König, D. (1985). Queueing networks: A survey of their random processes. *SIAM Review, 27*(3), 335–403.

Duenyas, I. (1994). Estimating the throughput of a cyclic assembly system. *International Journal of Production Research, 32*(6), 1403–1419.

Duenyas, I., & Hopp, W. J. (1990). Estimating variance of output from cyclic exponential queueing systems. *Queueing Systems: Theory and Application, 7*(3–4), 337–353.

Eager, D. L., Sorin, D. J., & Vernon, M. K. (2000). AMVA techniques for high service time variability. In *SIGMETRICS '00: Proceedings of the 2000 Association for Computing Machinery SIGMETRICS international conference on Measurement and modeling of computer systems* (pp. 217–228), Santa Clara.

Erlang, A. K. (1917). Solution of some problems in the theory of probabilities of significance in automatic telephone exchanges. *P.O. Electrical Engineers' Journal, 10*, 189–197.

Framinan, J. M., González, P. L., & Ruiz-Usano, R. (2003). The conwip production control system: Review and research issues. *Production Planning and Control, 14*(3), 255–265.

Frein, Y., & Dallery, Y. (1989). Analysis of cyclic queueing networks with finite buffers and blocking before service. *Performance Evaluation, 10*(3), 197–210.

Gaver, D. (1962). A waiting line with interrupted service, including priorities. *Journal of Royal Statistical Society, 24*(1), 73–90.

Gaver, D., & Shedler, G. S. (1973a). Approximate models for processor utilization in multiprogrammed computer systems. *SIAM Journal of Computing, 2*(3), 183–192.

Gaver, D., & Shedler, G. S. (1973b). Processor utilization in multiprogramming systems via diffution approximation. *Operations Research, 21*(2), 569–576.

Gelenbe, E. (1975). On approximate computer system models. *Journal of the Association for Computing Machinery, 22*(2), 261–269.

Gershwin, S. B., & Schor, J. E. (2000). Efficient algorithms for buffer space allocation. *Annals of Operations Research, 93*(1), 117–144.

Gonzales, E. A. (1997). *Optimal resource allocation in closed finite queuing networks with blocking after service*. Ph.D. thesis, University of Massachusetts, Amherst.

Gordon, W., & Newell, G. (1967). Closed queueing systems with exponential servers. *Operations Research, 15*(2), 254–265.

Helber, S. (2005). Analysis of flow lines with cox-2-distributed processing times and limited buffer capacity. *OR Spectrum, 27*(2–3), 221–242.

Helber, S., Schimmelpfeng, K., & Stolletz, R. (2011). Setting inventory levels of conwip flow lines via linear programming. *Business Research, 4*(1), 98–115.

Hopp, W. J., & Spearman, M. L. (1990). Conwip: A pull alternative to kanban. *International Journal of Production Research, 28*(5), 879–894.

Hopp, W. J., & Spearman, M. L. (2000). *Factory physics*. New York: Irwin McGraw-Hill.

Jackson, J. (1957). Networks of waiting lines. *Operations Research, 5*(4), 518–521.

Jackson, J. (1963). Jobshop-like queueing systems. *Management Science, 10*(1), 131–142.

Johnson, M., & Taaffe, M. (1989). Matching moments to phase distributions: Mixtures of erlang distributions of common order. *Stochastic Models, 5*(4), 711–743.

Johnson, M., & Taaffe, M. (1990a). Matching moments to phase distributions: Density function shapes. *Stochastic Models, 6*(2), 283–306.

Johnson, M., & Taaffe, M. (1990b). Matching moments to phase distributions: Nonlinear programming approaches. *Stochastic Models, 6*(2), 259–281.

Johnson, M., & Taaffe, M. (1991a). A graphical investigation of error bounds for moment-based queueing approximations. *Queueing Systems, 8*(1), 295–312.

Johnson, M., & Taaffe, M. (1991b). An investigation of phase-distribution moment-matching algorithms for use in queueing models. *Queueing Systems, 8*(1), 129–147.

Kleinrock, L. (1975). *Queueing systems. Volume I: Theory*. New York: Wiley.

Kobayashi, H. (1974). Application of the diffusion approximation to queueing networks, part 1: Equilibrium queue distributions. *Journal of the Association for Computing Machinery, 21*(2), 316–328.

Koenigsberg, E. (1982). Twenty five years of cyclic queues and closed queue networks: A review. *The Journal of the Operational Research Society, 33*(7), 605–619.

Koenigsberg, E., & Lam, R. C. (1976). Cyclic queue models of fleet operations. *Operations Research, 24*(3), 516–529.

Kouvatsos, D. D., & Almond, J. (1988). Maximum entropy two-station cyclic queues with multiple general servers. *Acta Informatica, 26*(3), 241–267.

Lean Enterprise Institute. (2003). *Lean lexicon: A graphical glossary for lean thinkers*. Brookline: Lean Enterprise Institute.

Lee, T.-E., & Seo, J.-W. (1998). Stochastic cyclic flow lines: Non-blocking, markovian models. *The Journal of the Operational Research Society, 49*(5), 537–548.

Li, J., Blumenfeld, D. E., Huang, N., & Alden, J. M. (2009). Throughput analysis of production systems: Recent advances and future topics. *International Journal of Production Research, 47*(14), 3823–3851.

Little, J. D. C. (1961). A proof of the ueuing formula: $L = \lambda$ W. *Operations Research, 9*(3), 383–387.

Liu, X.-G., Zhuang, L., & Buzacott, J. A. (1993). A decomposition method for throughput analysis of cyclic queues with production blocking. In R. O. Onvural & I. F. Akyildiz (Eds.), *Queueing networks with finite capacity* (pp. 253–266). Amsterdam: North-Holland.

Manitz, M. (2005). *Leistungsanalyse von Montagesystemen mit stochastischen Bearbeitungszeiten.* Ph.D. thesis, Universität zu Köln.

Manitz, M. (2008). Queueing-model based analysis of assembly lines with finite buffers and general service times. *Computers and Operations Research, 35*(8), 2520–2536.

Marie, R. A. (1979). An approximate analytical method for general queueing networks. *IEEE Transactions on Software Engineering, 5*(5), 530–538.

Marie, R. (1980). Calculating equilibrium probabilities for $\lambda(n)/c_k/1/n$ queues. In *PERFOR-MANCE '80: Proceedings of the 1980 International Symposium on Computer Performance Modelling, Measurement and Evaluation* (pp. 117–125), Toronto.

Marie, R. A., Snyder, P. M., & Stewart, W. J. (1982). Extensions and computational aspects of an iterative method. In *Proceedings of the 1982 Association for Computing Machinery SIGMETRICS Conference on Measurement and Modeling of Computer Systems* (pp. 186–194), Seattle.

Marie, R. A., & Stewart, W. J. (1977). A hybrid iterative-numerical method for the solution of a general queueing network. In *Proceedings of 3rd Symposium on Measuring, Modelling and Evaluating Computer Systems* (pp. 173–188), Bonn-Bad Godesberg.

Marie, R. A., & Stewart, W. J. (1983). On exact and approximate iterative methods for general queueing networks. *Publication Interne IRISA, 6*(24), 1–27.

Marshall, A. W., & Olkin, I. (2007). *Life distributions.* New York: Springer.

Matta, A., & Chefson, R. (2005). Formal properties of closed flow lines with limited buffer capacities and random processing times. In *The 2005 European Simulation and Modelling Conference.*

Mitzlaff, U. (1997). *Diffusionsapproximation von Warteschlangensystemen.* Ph.D. thesis, Technische Universität Clausthal.

Neuse, D., & Chandy, K. M. (1981). SCAT: A heuristic algorithm for queueing network models of computing systems. *Association for Computing Machinery Sigmetrics Performance Evaluation Review, 10*(3), 59–79.

Neuse, D., & Chandy, K. M. (1982).HAM: The heuristic aggregation method for solving general closed queueing network models of computer systems. *Association for Computing Machinery Sigmetrics Performance Evaluation Review, 11*(4), 195–212.

Neuts, M. F. (1982). Explicit steady-state solution to some elementary queueing models. *Operations Research, 30*(3), 480–489.

Onvural, R. (1990). Survey of closed queueing networks with blocking. *Association for Computing Machinery Computing Surveys, 22*(2), 83–121.

Onvural, R. O., & Perros, H. (1989a). Approximate throughput analyis of cyclic queueing networks with finite buffers. *IEEE Transactions on Software Engineering, 15*(6), 800–808.

Onvural, R. O., & Perros, H. (1989b). Some equivalencies between closed queueing networks with blocking. *Performance Evaluation, 9*(2), 111–118.

Osogami, T., & Harchol-Balter, M. (2006). Closed form solutions for mapping general distributions to quasi-minimal ph distributions. *Performance Evaluation, 63*(6), 524–552.

Osorio, C., & Bierlaire, M. (2009). An analytic finite capacity queueing network model capturing the propagation of congestion and blocking. *European Journal of Operational Research, 196*(3), 996–1007.

Perros, H. G., Nilsson, A., & Liu, Y. (1988). Approximate analysis of product-form type queueing networks with blocking and deadlock. *Performance Evaluation, 8*(1), 19–39.

Rall, B. (1998). *Analyse und Dimensionierung von Materialflußsystemen mittels geschlossener Warteschlangensysteme.* Ph.D. thesis, Technische Hochschule Karlsruhe.

Reiser, M. (1979). A queueing network analysis of computer communication networks with window flow control. *IEEE Transactions on Communications, 27*(8), 1199–1207.

Reiser, M. (1981). Mean-value analysis and convolution method for queue-dependent servers in closed queueing networks. *Performance Evaluation, 1*(1), 7–18.

Reiser, M., & Kobayashi, H. (1974). Accuracy of the diffusion approximation for some queueing systems. *IBM Journal of Research and Development, 18*(2), 110–124.

Reiser, M., & Lavenberg, S. S. (1980). Mean-value analysis of closed multichain queueing networks. *Journal of the Association for Computing Machinery, 27*(2), 313–322.

Ross, S. M. (1996). *Stochastic processes* (2nd ed.). New York: Wiley.

Ross, S. M. (1997). *Introduction to probability models* (6th ed.). New York: Academic.

Saad, Y., & Schultz, M. H. (1986). GMRES: A generalized minimal residual algorithm for solving nonsymmetric linear sytems. *SIAM Journal of Scientific Computing, 7*(3), 856–869.

Satyam, K., & Krishnamurthy, A. (2008). Performance evaluation of a multi-product system under conwip control. *IIE Transactions, 40*(3), 252–264.

Sauer, C., & Chandy, K. M. (1975). Approximate analysis of central server models. *IBM Journal of Research and Development, 19*(3), 303–313.

Sauer, C., & Chandy, K. M. (1981). *Computer systems performance modelling*. Englewood Cliffs: Prentice Hall.

Schassberger, R., & Daduna, H. (1983). The time for a round trip in a cycle of exponential queues. *Journal of the Association for Computing Machinery, 30*(1), 146–150.

Schmidt, R. (1997). An approximate mva algorithm for exponential, class-dependent multiple servers. *Performance Evaluation, 29*(4), 245–254.

Schweitzer, P. (1979). Approximate analysis of multiclass closed network of queues. In *Proceedings of International Conference on Stochastic Control and Optimization* (pp. 25–29), Amsterdam.

Schweitzer, P. J., Seidmann, A., & Shalev-Oren, S. (1986). The correction terms in approximate mean value analysis. *Operations Research Letters, 4*(5), 197–200.

Schwerer, E., & van Mieghem, J. A. (1994). Brownian models of closed queueing networks: Explicit solutions for balanced three-station systems. *The Annals of Applied Probability, 4*(2), 448–477.

Sevcik, K., & Mitrani, I. (1981). The distribution of queuing network states at input and output instants. *Journal of the Association for Computing Machinery, 28*(2), 358–371.

Shanthikumar, J. G., & Gocmen, M. (1983). Heuristic analysis of closed queueing networks. *International Journal of Production Research, 21*(5), 675–690.

Shum, A. W. (1980). *Queueing models for computer systems with general service time distributions*. New York: Garland Publishing.

Shum, A., & Buzen, J. (1977). The EPF techniques: A method for obtaining approximate solutions to closed queueing networks with general service times. In H. Beiler & E. Gelenbe (Eds.), *Measuring, modelling and evaluationg computer systems* (pp. 201–220). Amsterdam: North Holland.

Spearman, M. L. (1991). An analytic congestion model for closed production systems with ifr processing times. *Management Science, 37*(8), 1015–1029.

Stewart, W. J. (1994). *Introduction to the numerical solution of Markov chains*. Princeton: Princeton University Press.

Stewart, W. J. (2009). *Probability, Markov chains, queues, and simulation*. Princeton: Princeton University Press.

Strelen, J. (1989). A generalization of mean value analysisto higher moments: Moment analysis. *Association for Computing Machinery Sigmetrics Performance Evaluation Review, 14*(1), 129–140.

Sun, F. (2006). *Stochastic analyses arising from an new approach for closed queueing networks*. Ph.D. thesis, Texas A&M University.

Suri, R., & Desiraju, R. (1997). Performance analysis of flexible maufacturing systems with a single discrete material-handling device. *The International Journal of Flexible Manufacturing Systems, 9*(3), 223–249.

Suri, R., & Diehl, G. W. (1986). A variable buffer-size model and its use in analyzing closed queueing networks with blocking. *Management Science, 32*(2), 206–224.

Suri, R., & Hildebrant, R. R. (1984). Modelling flexible manufacturing systems using mean-value analysis. *Journal of Manufacturing Systems, 3*(1), 27–38.

Tempelmeier, H., & Kuhn, H. (1993). *Flexible manufacturing systems*. New York: Wiley.

Tempelmeier, H., Kuhn, H., & Tetzlaff, U. (1989). Performance evaluation of flexible manufacturing systems with blocking. *International Journal of Production Research, 27*(11), 1963–1979.

Tijms, H. C. (2003). *A first course in stochastic models*. New York: Wiley.

Tolio, T., & Gershwin, S. B. (1998). Throughput estimation in cyclic queueing networks with blocking. *Annals of Operations Research, 79*(0), 207–229.

Werner, H. (2010). *Supply chain management: Grundlagen, Strategien, Instrumente und Controlling*. Wiesbaden: Gabler.

Whitt, W. (1984). Open and closed networks of queues. *AT&T Bell Laboratories Technical Journal, 63*(9), 1911–1979.

Yao, D. D. (1985). Some properties of the throughput function of closed networks of queues. *Operations Research Letters, 3*(6), 313–317.

Yao, D. D., & Buzacott, J. A. (1985). Modeling a class of state-dependent routing in flexible manufacturing systems. *Annals of Operations Research, 3*(3), 153–167.

Yao, D. D., & Buzacott, J. A. (1986a). The exponentialization approach to flexible manufacturing system models with general processing times. *European Journal of Operational Research, 24*(3), 410–416.

Yao, D. D., & Buzacott, J. A. (1986b). Models of flexible manufacturing systems with limited local buffers. *International Journal of Production Research, 24*(1), 107–117.

Yüzükirmizi, M. (2005). *Finite closed queueing networks with multiple servers and multiple chains*. Ph.D. thesis, University of Massachusetts Amherst.

Yüzükirmizi, M. (2006). Performance evaluation of closed queueing networks with limited capacities. *Turkish Journal of Engineering and Enviromental Science, 30*(5), 269–286.

Zhuang, L., Buzacott, J. A., & Liu, X.-G. (1994). Approximate mean value performance analysis of cyclic queueing networks with production blocking. *Queueing Systems, 16*(1), 139–165.

Zhuang, L., & Hindi, K. S. (1990). Mean value analysis for multiclass closed queueing network models of flexible manufacturing systems with limited buffers. *European Journal of Operational Research, 46*(3), 366–379.

Zhuang, L., & Hindi, K. S. (1991). Approximate MVA for closed queueing network models of FMS with a block-and-wait mechanism. *Computers and Industrial Engineering, 20*(1), 35–44.

Zhuang, L., & Hindi, K. S. (1993). Approximate decomposition for closed queueing network models of fmss with a block-and-wait and state-dependent routing mechanism. *European Journal of Operational Research, 67*(3), 373–386.

Index

Arrival Theorem, 18, 22

Blocking, 8, 10, 75, 76, 121, 133

Closed queueing networks, 5
Convolution, 21, 23, 25, 26, 36
CONWIP, 7, 16

Deadlock, 9
Decomposition, 43, 45
Diffusion approximation, 33

EBOTT, 42
Exponential distribution, 73, 75

FMS, 16, 41

Global balance equations, 25, 91
GMRES, 116
Gordon-Newell Theorem, 19

Inter-departure-time distribution, 131

Little's Law, 17, 22, 122

Marie's method, 35, 38, 43
Markov chains, 26, 61, 73
Markov property, 64, 91, 92, 104
Mean value analysis, 17, 22, 23, 25, 26, 29, 31, 37, 38, 43

Non product form networks, 24
Normalizing constant, 20
Norton's Theorem, 18

Performance measures, 120
Phase-type distribution, 7, 25, 38, 43, 64, 73, 76
Product form networks, 19
Production rate, 11
Production rate function, 11

Starving, 10, 121, 123, 133
State space, 79

Theorem
    arrival, 18, 22
    gordon-Newell, 19
    little's Law, 17, 22, 122
    norton's, 18
Transition rates, 91, 111

S. Lagershausen, *Performance Analysis of Closed Queueing Networks*, Lecture Notes in Economics and Mathematical Systems 663, DOI 10.1007/978-3-642-32214-3, © Springer-Verlag Berlin Heidelberg 2013